Selected Issues in Mathematics Education

THE NATIONAL SOCIETY
FOR THE STUDY OF EDUCATION

Series on Contemporary Educational Issues
Kenneth J. Rehage, Series Editor

The 1981 Titles

Psychology and Education: The State of the Union, Frank H. Farley and Neal J. Gordon, editors
Selected Issues in Mathematics Education, Mary Montgomery Lindquist, editor

The National Society for the Study of Education also publishes Yearbooks which are distributed by the University of Chicago Press. Inquiries regarding all publications of the Society, as well as inquiries about membership in the Society, may be addressed to the Secretary-Treasurer, 5835 Kimbark Avenue, Chicago, IL 60637. Membership in the Society is open to any who are interested in promoting the investigation and discussion of educational questions.

Selected Issues in Mathematics Education

□

Edited by

Mary Montgomery Lindquist

National College of Education
Evanston

McCutchan Publishing Corporation
2526 Grove Street
Berkeley, California 94704

ISBN 0-8211-1114-0
Library of Congress Catalog Card Number 80-82903

Printed in the United States of America

Cover design by Terry Down, Griffin Graphics

Series Foreword

Developments in the field of mathematics education during the past two decades made it seem especially appropriate that the series on Contemporary Educational Issues include a volume on issues related to the teaching of mathematics. The recent appearance of *An Agenda for Action: Recommendations for School Mathematics of the 1980s*, published by the National Council of Teachers of Mathematics, confirms the timeliness of this volume on *Selected Issues in Mathematics Education.*

That this volume should be developed under the sponsorship of the committee on the Expanded Publication Program of the National Society for the Study of Education was suggested in the fall of 1977 by Professor Elizabeth Fennema, a member of that committee. The idea was explored further at a subsequent meeting attended by representatives of the Society's committee and of the publications committee of the National Council of Teachers of Mathematics. Following that meeting, Dr. Mary M. Lindquist, a member of the faculty of the National College of Education in Evanston, Illinois, was asked to prepare a detailed prospectus for the volume. The proposal was approved by the Society's committee and Dr. Lindquist agreed to assume the responsibility of editing the book.

The National Society for the Study of Education is most grateful to Dr. Lindquist for her untiring work on the manuscripts for this volume and to the several specialists in the teaching of mathematics who contributed chapters. We believe the book will merit the attention of all who desire to be informed about some of the most pertinent issues in this very important area of education.

Kenneth J. Rehage

for the Committee on the Expanded Publication Program of the National Society for the Study of Education

Contributors

Thomas P. Carpenter, University of Wisconsin at Madison
Mary Kay Corbitt, University of Kansas
Jane Donnelly Gawronski, San Diego County Department of Education
Douglas A. Grouws, University of Missouri at Columbia
Shirley A. Hill, University of Missouri at Kansas City
Henry S. Kepner, Jr., University of Wisconsin at Milwaukee
Thomas E. Kieren, University of Alberta
Frank K. Lester, Jr., Indiana University
Mary Montgomery Lindquist, National College of Education
Douglas B. McLeod, San Diego State University
Phares G. O'Daffer, Illinois State University
Alan R. Osborne, The Ohio State University
Thomas R. Post, University of Minnesota
Laurie Hart Reyes, University of Wisconsin at Madison
Robert E. Reys, University of Missouri at Columbia
Thomas E. Rowan, Montgomery County Public School
Ann K. Schonberger, Bangor Community College
Ross Taylor, Minneapolis Public Schools
Paul R. Trafton, National College of Education

Contents

Introduction

We are in the midst of an exciting time in mathematics education, though many mathematics educators would not agree that *exciting* is the word to describe the issues we are facing: the back-to-basics movement, minimal competencies, declining test scores, limited financial support, and so forth. These are not unique to mathematics; many other disciplines face these or similar issues. In this book we, as members of the mathematics education community, would like to share with the broader educational community as well as with our colleagues in mathematics education some perspectives on selected issues. Although these issues are addressed specifically to mathematics education, some are readily generalized to other disciplines, and all interact with the school as a whole.

With these common problems or concerns of education, why is this an exciting time in mathematics education? It seems to me that mathematics education has arrived at a maturing time. The National Advisory Committee on Mathematical Education (NACOME) report (1975) stated that:

> . . . mathematics education today is much like a healthy but not untroubled teenager. Its preteen period of enthusiastic preoccupation with the relatively clear-cut issues of content development has given way to a confused recognition of additional and much more elusive problems.

Mathematics education today is the healthier because it has lost a certain simplistic view of its educational challenges and an undue certainty about its answers. (p. 149)

That was five years ago. How quickly does the youthful mathematics education community mature? Certainly we have not completely matured. But it is this process of maturation that I find exciting. Seeing the new technology around us and how mathematics educators wrestle with the issues technology brings is exciting. Another source of excitement is to watch how mathematics educators face the question of sex-related differences in mathematics and take positive steps to open the study fully to both sexes and to see the mathematics educational research community work more closely together on common problems while reaching out for a broader base. There is challenge in seeing that we have not forgotten there are content questions yet to be answered and realizing that they fit into a larger picture. And finally it is exciting to see the mathematics education community looking at broader educational problems and not just at mathematical programs.

Not every possible issue is addressed in this book; from conception its scope was limited. First, it addresses only those issues concerning elementary or secondary school education. Thus, issues pertaining solely to preschool, to teacher education, to college, or to the growing population of adults seeking further education have not been included. However, many of these groups are touched tangentially. For example, the chapter on attitudes and mathematics certainly relates to all populations. The chapter on teacher variables has direct application to teacher education. The chapter on manipulative materials is relevant to preschool. The recommendations of the chapters on content could influence college curricula.

This book does not address issues to which mathematics education makes no unique contribution, such as broad teaching strategies, classroom management, and other general topics. It also does not address issues that could not be adequately covered in this format or that are already fully covered in other recent publications. Evaluation and research are two topics that were omitted because of these criteria. On the other hand, it does address some of those issues that NACOME described as presenting elusive problems, and it does address content questions because they are no longer clear cut.

Looking at the last twenty-five years in mathematics education, we see great activity—true of all education during this time. There are several excellent sources that trace developments in this period: the

Thirty-second Yearbook of the National Council of Teachers of Mathematics, *A History of Mathematics Education in the United States and Canada* (NCTM 1970) and the NACOME report (1975) are particularly helpful. Only a few highlights are mentioned here. The late 1950s brought a revolution—one often equated with the modern mathematics movement or the "new math." Beginning at this time and lasting for about fifteen years, the mathematics curriculum was influenced by the mathematics community itself. It was a time of great prestige for mathematics, of good financial support, curriculum innovations, emphasis on retraining of teachers, and an expansion of research.

During this period, we saw many recommendations for content changes—elementary arithmetic should be expanded to elementary mathematics by including geometry, probability, and statistics; secondary mathematics should include topics that had traditionally been reserved for college. We saw hopes that unifying concepts would be integrated into the curriculum for the purpose of deepening the understanding of mathematics, and saw attempts to clarify language and symbols. All of these, as well as other goals of the modern mathematics period, were laudable goals that influenced the mathematics curriculum.

Whenever there are changes proposed or made, as there were during this period, awareness is heightened and everything becomes subject to scrutiny. At some point, an oversimplified view was taken, and many problems were blamed on the new math. The NACOME report (1975) takes a rational stand on the outcomes of this period:

> From a 1975 perspective the principal thrust of change in school mathematics remains fundamentally sound, though actual impact has been modest relative to expectations.
>
> The content innovations K–12, the emphasis on student understanding of mathematical methods, the judicious use of powerful unifying concepts and structures, and the increased precision of mathematical expression have made substantial improvement in the school mathematics program. Unfortunately, the innovations have not fulfilled the euphoric promise of 1960, [and] the current debate seems intent on locating blame for failures in real or imagined "new math" programs. . . . Furthermore, the acrimonious criticism sends many teachers and laymen in retreat to truly outmoded curricular goals, rather than moving forward by building on positive features. (p. 21)

At the same time that this so-called revolution was occurring, there was also another one. As Weaver and Suydam (1972) state:

> But there has been another revolution of import—a revolution which influenced elementary-school mathematics programs and instruction in particular. It is a revolution which all too often has been lost in the shadow of the "modern mathematics" movement. We refer to "The Revolution in Arithmetic" characterized by Brownell (1954) in the first article of the first issue of the *Arithmetic Teacher*. This "other revolution" was more rooted in educational psychology than in mathematics per se. But it had an appreciable impact upon the goals and content of elementary-school mathematics programs as well as upon the process of instruction. (p. 1)

The authors then proceed to describe the "meaning theory" of arithmetic as espoused by Brownell.

The focus of this "revolution" was how to make mathematics learning and instruction meaningful to all students. This was not at all contradictory to the goals of the new-math revolution, but several groups—teachers, curriculum developers, and some mathematics educators—saw more need to concentrate in this area. Therefore, after the wave of new-math materials, child-centered programs, activity approaches, and individualized programs began to appear. There was not just one thrust but many different ones, and it was a quieter, more subtle revolution, if indeed it could be called a revolution.

By the middle of the 1970s the back-to-basics movement was predominant, replacing both of the previous "revolutions." Curriculum became more skill oriented, and rote learning was given greater emphasis. Mathematics was no longer in a period of enchantment nor was there a great deal of public support.

This movement back to the basics is not over; but as we enter the 1980s, a balance is returning to the scene. Mathematics educators have taken a broad view of what is basic: there is a call for improving skills in problem solving and in the development of concepts, the neglected students are on our minds, the need for careful instruction is recognized, and the need to adjust to our changing society is being addressed. While we have not solved all our problems, we do have a sense of where we need to place our energies.

This book examines some of these places and gives guidance useful in making decisions. If we are thoughtful and purposeful, we should be able to learn the lessons of the past twenty-five years, culling the best for the decade to come.

REFERENCES

National Advisory Committee on Mathematical Education (NACOME). *Overview and Analysis of School Mathematics*. Washington, D.C.: Conference Board of the Mathematical Sciences, 1975.

National Council of Teachers of Mathematics (NCTM). *A History of Mathematics Education in the United States and Canada*. Thirty-second Yearbook of the National Council of Teachers of Mathematics. Reston, Va.: The Council, 1970.

Weaver, J. Fred, and Suydam, Marilyn N. *Meaningful Instruction in Mathematics Education*. Columbus, Ohio: Educational Resources Information Center (ERIC), June 1972.

PART ONE
Mathematics Curriculum

1. Assessing the Mathematics Curriculum Today

Paul R. Trafton

Mathematics occupies a well-established position in the school curriculum. While there is wide acceptance of the importance of mathematics, there is also a lack of consensus regarding the content of the curriculum, how that content should be treated, and the overall purposes for the study of mathematics. The public expresses concern about the mathematical competence of students, which causes educators to examine the achievement of students more closely. Many curriculum generalists are concerned about the fit between knowledge of subject matter and their view of the overall purposes of schooling. Mathematics educators are concerned about the narrow focus of many school programs, the limited treatment of content that they view as important for students, and the lack of attention given to developing proficiency in reasoning and problem solving.

The central purpose of this chapter is to assess the mathematics curriculum of today. It is important to examine that curriculum periodically. While issues are never fully resolved and critical questions never completely answered, a thoughtful examination of some of the problems and questions at a given period in time can sharpen the focus of discussions and provide a beginning point for making adjustments. The present educational climate is not ideal for the reflective deliberation and broad-based discussion necessary for the process to lead to substantive adjustments and change. Nonetheless it is still important to face the questions, begin making changes, and address the existing needs in order to move toward the goal of a more appropriate and productive mathematics program for our students.

The audience for this chapter is a broad one. Classroom teachers, curriculum generalists, and school administrators all look for guidance in understanding the mathematics curriculum and in knowing how to address current curriculum questions. The mathematics curriculum poses problems for many educators, and the overall intent of general education goals in mathematics education are not clear. Most educators have neither strong backgrounds in mathematics nor close links with the mathematics education community. Mathematics educators, particularly at the elementary level, come from a different background and set of traditions than many other educators, which increases the problem of communication. The prevailing instructional practice in mathematics sometimes concerns educators who see an inflexibility in the way the subject is taught and a lack of sensitivity in adjusting content goals and instructional approaches to the needs of students. In addition, a workable mathematics curriculum seems to require conditions that are sometimes in conflict with broad conceptions of curriculum and instruction.

This chapter consists of two sections. First, some facts that need consideration when viewing the curriculum and making curriculum decisions are discussed. Then five aspects of the curriculum that need particular attention today are examined. In each area adjustments need to be made if imbalances are to be corrected and problems resolved. However, specific changes in the mathematics curriculum need to be made in the context of the parameters for effective programs that is presented in the first section.

BASIC CURRICULAR CONSIDERATIONS

Discussions of curriculum cannot be conducted in a vacuum, nor should they comprise broad speculations and armchair observations. Rather they need to take into account some overall perceptions regarding the curriculum and to reflect some of the variables that influence the effectiveness of school programs. An appropriate curriculum that works well in the classroom is a delicate balance of many factors and a variety of considerations. There are four basic considerations that should undergird curriculum planning.

Mathematical Needs of the Students

Curriculum planning must take into consideration the final outcomes of the study of mathematics based on students' long-term needs for mathematics. We know the need for arithmetic to govern one's personal affairs and the requirements of many vocations using only a modest amount of mathematics. Two other components are not as well recognized. The first is the mathematical literacy component of the general education goals in mathematics, including the level of mathematical knowledge needed to comprehend and interpret events in the world as well as that which well-educated individuals should possess. Even a casual reading of newspapers and periodicals reveals much mathematics that is part of current issues and general news. Many aspects of good consumer awareness require a fairly sophisticated knowledge and understanding of mathematics.

The second component is the specialized training in mathematics that is prerequisite for study in a wide range of fields. Secondary school students often choose not to complete a full course of study, and they are not made aware of the implications of this choice. When these students investigate career options later, they frequently find that they face roadblocks or closed doors because of an inadequate background in mathematics. A careful study of a large number of careers and their mathematical requirements leads one to see the importance of the study of substantial amounts of mathematics by all students.

It seems essential, therefore, that curricular considerations focus carefully on questions of what content knowledge and proficiencies the majority of students should possess as a result of their study of mathematics. Discussions of the elementary and middle school curriculum frequently seem to assume that only social uses of mathematics need to be considered. At the secondary level the curriculum for students with weak backgrounds reflects a terminal orientation largely limited to computational arithmetic. Determining the mathematical content, skills, and applications that should constitute the curriculum is a far more complex and substantive process than is often recognized.

Organization of the Curriculum

There are many ways of approaching curriculum organization. Within subject matter fields, the curriculum can be broken into units of study that have little relationship to each other, as is often done in schools with the social studies curriculum. The curriculum can also be organized around an exhaustive list of discrete skills. At times the content can be handled within integrated units of study in which subject matter lines are blurred and content from various fields is taught if needed to fulfill the goals of the unit.

Each of these approaches has been attempted with the mathematics curriculum with less than successful results. Examining the content of mathematics and how specific mathematical content is learned supports a curriculum in which the content is carefully organized, carefully sequenced, and systematically taught. That is, a curriculum that has coherence and structure is suited to the structure inherent in the discipline of mathematics and the requirements for producing stable learning within students.

Mathematics consists of several strands of knowledge and these strands—such as numbers and operations, geometry, and statistics—should be the focus of the curriculum. It is productive to build these notions throughout the curriculum so that the student's understanding is regularly reinforced and expanded. The skills of mathematics can be linked to these major content strands. The development of the skills themselves also requires systematic study, careful sequencing, and extended development if they are to be well learned. While the systematic study of mathematical content and the sequential structuring of the curriculum can be extended to extremes, they are important to a sound curriculum.

The need for a coherent and structured curriculum does not mean there is no room for integrating mathematics with other areas of knowledge in appropriate ways, nor that all content has the same sequencing needs. It also does not mean that identifying specific outcomes should be neglected. It does suggest that the concerns expressed by mathematics educators are well founded when other approaches are advocated as the main organizing principle of the curriculum.

Instruction and Learning Input

Curriculum discussions generally focus on the content to be taught, the goals of the curriculum, and the organization of the curriculum content. Learning and instruction are two other variables that are included in examining mathematics education. While these three elements are often treated somewhat independently, they should be viewed as being dynamically interrelated. Examination of each of these elements needs to be conducted in the awareness that each is influenced by the others and there is considerable overlap between them.

It is important to recognize that curriculum discussions need to take into account knowledge about components of effective instruction and information about the learning of mathematics. It sometimes seems that curriculum is discussed in isolation from these other elements. The assumption appears to be made that students will learn whatever is taught to them. While it appears true that what is not taught will generally not be learned in mathematics, it is not true that just because something is taught it will be learned. Investigations of students' learning of mathematics provides valuable input regarding what content can appropriately be presented at a given level, the amount of time needed to successfully achieve the goals, and the kind of instruction required. Without this input, it is possible to make unwise decisions about the content to be included and to overestimate what can profitably be included in the curriculum. As one reflects back on the curriculum reform efforts of the early 1960s, there was not an adequate knowledge base about the ability of students to learn the content nor about the kind of instruction required. The reform movement provides strong evidence that while idealism and good intentions are desirable, they are not sufficient for effective curriculum planning. Curriculum thinking should always explore uncharted waters, but it should also draw insightfully on the knowledge provided by thoughtful investigations of instruction and learning.

Continuity and Change in Curriculum Development

Curriculum work is a never-ending process. There needs to be ongoing assessment of the content, ways of treating the content, and the effects of the curriculum on students. It is a process that takes place at several levels. It occurs as mathematics educators speak out on curriculum questions and publish in professional journals. It occurs as professional organizations, task forces, and curriculum groups on a national or regional level deal with curriculum issues. It also occurs at local levels where curriculum guides are produced that indicate priorities and make adjustments in standard curriculum materials.

Curriculum decision making needs to reflect the complementary roles of continuity and change. Examination of curriculum efforts over the years and knowledge about the psychology of human behavior suggest that sustained, productive progress is the result of dealing with both continuity and change appropriately. Maintaining continuity enables new efforts to be built on present aspects of the curriculum that possess strength and worth. Continuity serves as a check on hastily conceived change whose long-term effects have not been assessed. It provides a way of dealing with change in a manageable fashion, and it reflects the pragmatism inherent in American education.

Yet continuity is only one aspect of ongoing curriculum progress. A curriculum that meets the long-term needs of students must also be in a state of change. The content and skills that students need in order to meet technological and social demands change over time. New evidence is constantly available about better ways of teaching and sequencing the content of the curriculum. Thus change is essential in meeting new challenges and preserving us from the unquestioned support of the status quo, as well as from the regressive tendencies of not having a forward look to the curriculum. The idea of change also feeds the idealism that is essential to educational thinking and planning.

Curriculum change is inevitable and desirable. Yet the path to progress that translates into daily programs in individual classrooms is a complex one. Among the many variables to be considered is how to deal successfully with the need for both continuity and change.

ISSUES IN THE SCHOOL CURRICULUM

It is difficult to accurately identify *the* curriculum issues that exist at any one time. The identified issues reflect an individual's perspective on the curriculum and vantage point for viewing the curriculum. In this section several pervasive issues that affect school mathematics on a broad scale are identified and discussed. Each area discussed poses genuine problems, yet in each case there are workable ways of making adjustments.

The Treatment of Computation

Any discussion of the school curriculum must deal realistically and thoughtfully with computation. To many people the mathematics curriculum in kindergarten through grade eight is synonymous with computation. And in a broader context, the mathematics curriculum at all levels tends to be defined in terms of the many skills that are part of the subject. While mathematics educators reject this limited viewpoint, it is one that permeates the thinking of some educators and the public.

Computation has been a focal point in curriculum thinking over the past fifty years in mathematics, both in providing the impetus for change and in defining the nature of change. Today computation is again exerting great influence on the curriculum. Reasons for this include the traditional view of the importance of computation, the vacuum resulting from the rejection of the content and emphasis of the reform efforts of the 1960s, the cluster of factors that gave rise to and support the basic skills movement, the lack of a clear, well-articulated direction for the mathematics curriculum that has wide acceptance, and the lack of confidence teachers feel in dealing with other content.

It is difficult to achieve a balanced viewpoint on the role of computation in the curriculum. The topic is often not well served in discussions by its supporters or detractors. Many times its importance is either overstated by the public or underplayed by curriculum reformers. At present it is clear that computation dominates the curriculum that students experience in the classroom. Few would deny that computation (and mathematical skills, in general) is an important component of mathematical learning and necessary for any future uses of mathematics. Thus the key question is not whether computation is important or not. Rather, questions should center on how computation should be treated in the curriculum and what constitutes appropriate computational proficiency at this time. These questions are addressed in the following sections.

Computation and the Arithmetic Curriculum. In recent years computation frequently has been equated with arithmetic, although it constitutes only a portion of arithmetic. Arithmetic also includes a large array of number concepts, understandings and relationships with several sets of numbers, properties of the numeration system, meaning of operations on sets of numbers, and uses of number ideas and computation in real-world settings. Thus to equate computation with arithmetic clouds the discussion, as well as oversimplifies the complexity of the necessary work with numbers and operations. On one hand equating computation and arithmetic tends to remove the study of computation from the rich context in which it must occur, while on the other hand it leads to the inaccurate judgment that the elementary mathematics curriculum in conception consists primarily of computation. While that may be the case in many classrooms, it is not the intent of the curriculum.

Two major criticisms can be raised about the treatment of computation. First, computation is being taught out of the context of arithmetic concepts, relationships, and applications; and second, there is an undue reliance on drill to attain computational goals. While neither of these problems is new, it is disappointing that current educational practice does not reflect what is known about effective work in this area.

It is productive from both a curricular and learning point of view to examine the relationship between concepts, skills, and applications. These three areas are not only closely related, but they are mutually supporting as well (Figure 1–1). Computational skill is heavily influenced by major ideas of the operations and place value. Conversely, proficiency with computation is often a factor in comprehending major mathematical concepts. A similar interrelationship exists for skills and applications. A strong use of applications gives purpose and significance to computation. In addition, linking applications to computation throughout makes the skills more readily available when needed to solve problems. A curriculum that regularly integrates these elements is necessary to achieve the broad goals of a basic mathematics education as well as to attain the necessary computational objectives. To put it another way, evidence suggests that computation in isolation reduces the probability that computation will be well learned and makes it more difficult to attain goals for the further study of arithmetic.

The use of drill as the primary device to develop computational proficiency can also be criticized. Investigations of learning basic facts and algorithms indicate that meaningful, well-sequenced developments are effective in developing skills, along with a reasonable amount of drill. Much can be written on this topic from a learning perspective, but it will suffice here to note that inadequate teaching of computation greatly expands the amount of time needed to teach computation, thus making less time available to teach other content and attain other goals; this in turn causes the curriculum to stay in a "holding pattern" from year to year.

Appropriate Computational Proficiency. The level of computational proficiency deemed necessary for students needs careful consideration, particularly with the widespread availability of calculators. While most curriculum materials have kept the level within reasonable limits for several decades, the extent to which computation is carried is far greater in many classrooms and in some drill materials. Furthermore, current pressures are causing an increase in the level of computational work. Certainly students need to be able to compute, but it is difficult to justify a strong emphasis on addition with several five- or six-digit addends, on multiplication where both factors are three or

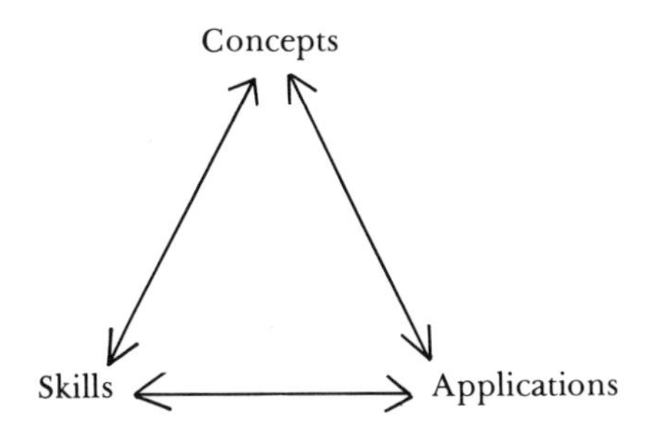

Figure 1–1

The interrelationship of skills, concepts, and applications

more digits, on long division where the divisor or quotient each have three or more digits, or on computation involving seldom-used fractions. We need to adopt reasonable limits for the teaching of computation. While computation can and should be taught well, advanced levels of computation pose great problems for many students and consume an unjustifiable amount of curriculum time.

Curriculum Content

The second major issue that effects all students directly deals with the mathematical content included in the curriculum. While this issue has obvious relevance in grades kindergarten through eight, it applies as well to the secondary level for students whose needs require an alternative to the standard curriculum. Mathematics educators have long advocated a broadly oriented curriculum that includes content from geometry, measurement, statistics and graphing, probability, functions, and number theory. It is further argued that the content be presented so as to stress reasoning and problem solving, alertness to the reasonableness of results, and estimation.

A case for a curriculum with expanded content needs to be carefully established. The reasons need to be understood and accepted by classroom teachers, school administrators, and the general public if the curriculum that students experience is to reflect such content. Clearly the reasons need to go beyond the fact that the content is "good mathematics" or that students will enjoy it.

The learning of useful knowledge has long been accepted as a major goal of education. Thus the case for a comprehensive mathematics curriculum needs to be constructed on the base of useful knowledge if it is to be accepted by educators and the public. In an earlier section it was argued that there is a strong need to accept a broader conception of what constitutes a basic mathematics education. Certainly school mathematics needs to prepare students for dealing with mathematics that they will encounter in the diverse situations of daily life. It also needs to provide them with a knowledge base that enables them to study additional mathematics successfully and to qualify for various careers.

It is relatively easy to accept, in general, a curriculum with a broad scope. The task becomes more complex as individual topics are examined. Many times, seeing the usefulness of particular content requires a relatively strong mathematics background and an awareness of the ways in which the content can be applied. One example of this is the area of geometry. A careful study of geometry indicates that there are many direct and indirect uses for a great deal of geometry, both vocationally and in daily life, and that geometry interacts with number ideas and algebraic notions in substantive ways throughout mathematics. Yet when most individuals think about geometry they have difficulty seeing its usefulness.

Mathematics educators need to recognize that a lot of pragmatic work needs to be done to prepare compelling reasons for including additional content. The "obviousness" of the content's importance is not always clear to others, and this importance can be difficult to establish at times, particularly in terms of the inclusion of content X at point Y in the curriculum. In addition, it needs to be made relatively clear what it is that students need to know about the content and what they should be able to do. By and large, mathematics educators have not addressed themselves to this task effectively. General appeals and the restating of arguments based on premises already accepted by other mathematics educators do not successfully communicate to the diverse groups whose support is essential if change is to occur.

The case for a comprehensive curriculum can be supported in additional ways as well. First, the development of major mathematical ideas over an extended period of time leads to a better-articulated curriculum. Competence and confidence with content require continued exposure to the content, since understanding and insight develop over long periods of time. Many problems that students have in algebra and geometry, for example, would be reduced if some of the basic notions of these content areas were treated appropriately in earlier years. The concept of strands of content and processes, which was strongly stressed in the 1960s, is a valid one whose implementation would result in a curriculum that better fits the way students gain mastery of mathematics.

A second source of support for a comprehensive curriculum is the evidence from research on learning and instructional approaches. During the past several years we have learned much about the kinds of experiences appropriate for students at various levels and what they are able to learn well when appropriate instruction is provided. Thus we have increased knowledge about the fit between specific content and the ability of students to deal with this content.

Finally, a comprehensive curriculum makes many contributions in the affective area. The mathematics education of students can make them aware of the pervasiveness of mathematics and the many contributions mathematics has made to the development of civilization. At present, many adults have a narrow view of what constitutes mathematics and its usefulness, but a broad curriculum has the potential of raising perceptions in this area. It is also important that students learn to appreciate the value of mathematics and have confidence in doing mathematics. Work with the expanded content suggests that students do enjoy studying it and gain proficiency with it.

At the secondary level the traditional curriculum is well established. It is highly oriented toward the study of advanced mathematics and the in-depth study of particular branches of it. A case can be made at this level for establishing a broader curriculum that includes more work with probability and statistics, a more informal, wide-ranging treatment of geometry, and work on the uses and programming of computers. A multiyear, well-articulated program would be appropriate for the noncollege-bound students as well as for precollege students whose need for mathematics is more broad and general. Such a curriculum should include aspects of the traditional curriculum that are needed by most citizens. Past efforts at developing such a curriculum suggest that instituting this type of program is not easy. The need is there, however, and it needs to be addressed.

Programs for Low Achievers

A third issue of widespread importance is that of programs for low-achieving students. Two general observations can be made. First, there are large numbers of students for whom the learning of mathematics is unsuccessful. The longer they are in school, the wider the gap grows between what they know and what they should know. Second, most programs for low-achieving students do not work very well.

There is frustration on the part of those who teach these students and/or there is little growth in achievement. We often fail to recognize the depth of these students' problems and underestimate the kind of help they need. Frequently the help these students receive represents "more of the same"—rapid reteaching of isolated concepts and skills followed by large amounts of drill. The problem is particularly acute at the middle and secondary school levels where the frustration of the students is greater, the problems are more extensive, the diagnostic work is more complex, and the teachers lack the needed primary and intermediate grade background. Many times teachers are asked to teach large numbers of these students with no overall program design and only a single textbook from which to work.

It is possible, however, for these programs for low achievers to be successful. Much is known about the components of learning and the instruction of specific content that is directly applicable to low achievers, and there are existing programs to examine. These are some important aspects of providing viable programs.

1. We need to recognize that most of these students can learn, want to learn, and will learn when appropriate programs are provided. The fact that they have not yet learned is our problem as much as it is theirs. When students sense that they are in fact learning successfully, they exhibit positive attitudes and a willingness to work seriously at overcoming their deficits.
2. The content needs to be carefully selected. For many of these students the curriculum needs to focus on those concepts, skills, and applications that are critical for managing one's personal affairs and being employable in the marketplace. This often means difficult choices with respect to content selection, with the study of some desirable content being deferred. This also means deleting less central or more sophisticated aspects of many topics.
3. Instruction needs to begin at basic levels and proceed slowly enough so that the students have enough time to learn and integrate all the components of a task. The failure to go "low enough" and "slow enough" is a major cause for the ineffectiveness of many programs. The problems of remedial middle-school students in the area of numeration are often at the level of tasks that comprise work with three-place numbers. Instruction also needs to focus on students mastering whatever they study. Students strongly need a firm base on which to construct more advanced concepts and skills.
4. There needs to be a developmental approach to learning that focuses on the link between models and symbols, on careful sequencing of work, and on linking the skill work to conceptual

components. Focusing only on skill components hides the true source of the problems and does not lead to long-term learning.

5. Curriculum materials that are specifically designed for use with low achievers are needed. Building the necessary developmental work and sequencing into the materials ensures that students focus on the key elements of a learning task, and this in turn increases teacher effectiveness. While there is an abundance of drill materials, there are few learning materials appropriate for this audience.
6. Programs need to have a comprehensive design including well-thought-out assessment procedures, provisions for individualization, appropriate curriculum materials, maintenance procedures, and ways of assessing the overall program effectiveness. While these students can learn, strong gains require the careful fitting together of several components. It is unrealistic to expect individual teachers to be able to construct and integrate all of the pieces.
7. Programs for low achievers need strong administrative support. They need to be accepted as a top priority and given strong financial support. There has to be a willingness to make adjustments in other aspects of the school program, including scheduling, and to provide support and ongoing training for those teaching in the program.

The times in which we live require that all students master as much mathematics as possible. We cannot afford to have large numbers of people who have only a severely limited proficiency in mathematics. While low achievers and programs for low achievers pose difficult problems, we must apply what we know about effective instruction and sound learning. Only to claim that we did what we could while students who can learn fail to learn is not acceptable.

The Role of Applications

A historical study of the field of mathematics reveals that many mathematical ideas developed in response to real-world problems and also indicates that mathematics is applied in a variety of ways from simple social situations to the use of sophisticated mathematical models for analyzing situations drawn from all branches of science, including social science. Yet the mathematics curriculum is often criticized for being devoid of applications or treating them in a cursory, limited fashion. Certainly many adults view mathematics as a set of rigid, abstract rules and feel that it is not very useful. Older students frequently ask to have the usefulness of some content demonstrated. Unfortunately, many times this is not done, and many students simply do not find the study of mathematics satisfying for its own sake. One student relates questioning how the content could be applied throughout high school and into college. He was told that he would have to wait until the next year or another course, and his final response was to drop the study of mathematics, since he saw no use for it.

The call for more applications of all types in mathematics courses is not new, and it is difficult not to support it in theory. Yet translating theory into practice is always difficult, and there are many problems associated with the use of applications. It is often difficult to find appropriate, substantive applications that students can handle; applications need to be selected carefully if they are to be appealing to students. The use of applications requires a dual focus in teaching—the careful focus on developing concepts and skills, with the various complexities involved, and the focus on applying the content. Finally, it can be difficult to find time in the curriculum to devote to applications; teachers do not have extensive training in this area, and there are few curriculum materials available.

Nonetheless the present is an ideal time to make renewed efforts in this direction. Within the discipline of mathematics, applications have strong support. The pragmatic orientation of current thinking supports applications, and the widespread use of calculators makes it possible to handle the "messy" data and complex computation that is part of work with applications, thus freeing students to focus on the problem solving involved. While additional curriculum development is needed, there are already steps that can be taken by teachers at all levels to do more in this area. These steps include the following:

1. Use real-world situations to introduce new content. This helps motivate the learner to master the content as well as makes applications an integral part of the content itself. A deep, positive impression in my own experience was the first calculus class that used a problem about maximizing income through the pricing of circus tickets to indicate what calculus dealt with and where the course was headed.
2. Use applications, including word problems, on a daily basis so that students view them as a normal part of the study of mathematics and gain confidence in working with them.
3. Emphasize those areas of mathematics that are rich in applications. Two primary areas are measurement and statistics, includ-

ing graphing. Obviously the applications aspects should be maximized.

4. Set days aside for class and individual projects that deal with applications. Examples of this might include a series of measuring and orienteering experiences in the outdoors, projects in collecting and analyzing data, and working cooperatively with other departments at the middle-school level to investigate a large-scale topic in an interdisciplinary manner.
5. Build applied experiences with arithmetic into the secondary curriculum. At present, students get little maintenance work with the content or its applications once they enter the ninth grade. The current thrust on consumer mathematics is a highly appropriate one. Certainly high school students need regular contact with arithmetic, and they need to deal with more complex and extensive situations using arithmetic. Whether the need can best be met by a separate course taken in the eleventh or twelfth grade or by a series of short-term experiences in regular courses is not clear. What is clear, however, is that much more needs to be done at the secondary level.

Topic Placement in the Present Curriculum

In previous sections suggestions have been made for changes in the treatment of present curriculum content and the need for the inclusion of additional content. The recommendations were motivated by the need for a more appropriate curriculum in meeting long-term goals and a more workable curriculum on a daily basis. One other factor to be considered is adjustments in the grade placement of topics in the present curriculum. Efforts to teach particular content at inappropriate times result in ineffective instruction, poor achievement, and the use of large quantities of valuable instructional time. As a result, additional pressure is placed on teachers and students, as well as on the curriculum, and the curriculum tends to become more limited. There are two areas in kindergarten through grade eight where changes in the grade placement of current content would provide a more workable curriculum and create additional time for use in more productive ways, and there is one area at the secondary level where the scope and sequence needs to be reexamined.

Whole Numbers. In recent years a downward shift has occurred in the grade placement of computational topics. As a result, they are taught before most students can profitably deal with them. Not only are large amounts of time devoted to them, but the following year it is still necessary to spend much time redeveloping these topics. Some examples of this are:

1. Expectations for addition and subtraction facts at grades one and two and for multiplication and division facts at grade three. Students need much work with these facts over time before they can commit them to long-term memory and use them with confidence.
2. The two-digit subtraction algorithm involving renaming in grade two. Students appear to lack the maturity and the sound knowledge of prerequisite content to be able to master this algorithm in grade two.
3. The treatment of four-place numbers in grade three and larger numbers in grade four. Students have difficulty in comprehending large numbers and often spend much time with pointless exercises before they are capable of dealing with them insightfully.
4. Multiplication with two two-place factors for below-average students in grade four. This is a complex skill in which the level of processing is difficult and in which there is interference with other skills.
5. Long division with two-place divisors in grade four. While the work with one-place divisors can be learned at this level, it is a difficult skill and the process needs time to be fully mastered. The two-place divisor work exceeds the capability of most fourth-grade students.

Fractions and Decimals. The movement to the metric system and the use of calculators have caused a reexamination of the placement of fractions and decimal topics. This reexamination suggests a reduction in the amount of fraction work beginning in grade four with the treatment of decimal topics starting at that level. Both of these shifts are supported by research evidence, which suggests that many topics with fractions are taught before students are able to deal with them. While preliminary decimal work can begin in grade four, there is evidence to suggest that successful work with decimals to tenths can be done in grade three. Examination of students' future needs for fractions and decimals suggests that concepts and skills in both areas still need to be taught. While an extensive treatment of the scope and sequence of these topics is not possible here, some brief guidelines are presented to give some direction.

1. Fractions: It is recommended that work with the rule for equivalent fractions be delayed until grade five. Also, work with addition

of unlike fractions should be delayed until this time, with much of the work at this level limited to proper fractions or mixed numbers requiring limited renaming. There is also support for delaying work with multiplication and division of fractions until grade six for most students, reserving the more complex mixed number aspects until grade seven.

2. Decimals: The trend toward introducing decimal notation for tenths and hundredths in grade four, along with addition and subtraction, seems appropriate. In grade five, the place value aspects of decimal numerals should receive attention, along with preliminary work with multiplication of decimals. In grades six and seven, the remaining decimal concepts and computation can be presented. In general, such a sequence would be workable for students and fit their needs for using decimals.

The Scope and Sequence of Geometry. Geometry holds an entrenched position in the high school curriculum, with the major emphasis on proof. Concern is often expressed about the heavy emphasis on proof throughout the course as well as the preliminary work that precedes proof. There is also concern about the great difficulty many students face in the course. Another problem area that is not as well recognized is the geometry strand at the middle-school level and the lack of any well-developed and well-articulated approach to the geometry of the middle and secondary schools. Part of the problem students face in the conventional high school course is the lack of a strong background in geometric knowledge and intuition on which to build a more in-depth treatment of geometry.

The direction for the secondary school geometry course is uncertain. There are many suggestions for alternative ways of organizing the content, and there have even been attempts to build a geometry course in which proof does not play a central role. Some change within the basic high school geometry course is needed and we also need a coordinated approach to geometry in grades seven through twelve that builds geometric ideas carefully and better links them to algebraic notions as well.

SUMMARY

We have examined factors that need to be considered when thinking about the mathematics curriculum as well as major areas that need careful attention when examining that curriculum. On the whole, the current mathematics curriculum can be characterized as being in a holding pattern. The residue remaining from the curriculum movement of the 1960s and the experimentation with respect to instructional management and approaches to schooling of the 1970s, together with the basic skills movement of recent years, has created a climate that promotes a somewhat narrow view of the general education goals of mathematics education. Within mathematics education there are several areas of activity and various suggestions for change, but there is neither a cohesive, well-integrated position emerging from this work nor does there seem to be strong support within education or from the general public for change.

However, there are hopeful signs. The concern of the public for strong academic programs, the investigations on how students learn specific topics, the renewed interest in problem solving and applications, and the contributions that calculators can make are all factors providing possible beginning points for future curriculum considerations. The present curriculum as conceived has strengths that should not be ignored, and this should be recognized.

Thus there is the potential for positive curriculum development. For new efforts to have widespread impact, however, they must include several variables. The examination of previous efforts and the current educational scene suggest that any new directions must have the support of the public and the educational community, must fit the criteria of useful mathematical knowledge as interpreted broadly, must deal thoughtfully with considerations about child growth and development, must be supported by research on classroom learning of the content, must include careful work with the aspects of mathematics programs that the public presently values, must fit what teachers can reasonably implement in the classroom, and must be capable of being implemented without the massive expenditure of funds. While these factors place many constraints on curriculum change, attention to them when planning a mathematics curriculum increases the likelihood that potential change will become actual change.

PART TWO
Mathematics Content

2. Problem Solving: Is It a Problem?

Frank K. Lester, Jr.

Learning to solve problems is the principal reason for studying mathematics. (National Council of Supervisors of Mathematics 1977, 1978)

Anyone who has studied mathematics at any level knows that problem solving lies at the heart of doing mathematics. Indeed, there is support for the notion that the ultimate aim of learning mathematics is to be able to solve problems. Despite this well-recognized general importance, the role problem solving should play in school mathematics is less clear, which is due in large part to the complex nature of problem solving. There is in fact general agreement that problem solving is the most complex category in all human learning, so complex as almost to defy description and analysis.

The importance of problem solving in mathematics is attested to by the innumerable books, monographs, and journal articles devoted to it. In addition, problem solving has been the major focus of several conference reports and curriculum development efforts. The complexity of this area of mathematical activity has caused it to be the subject of more research in recent years than any other topic in the mathematics curriculum.

To assist in establishing the problems and issues associated with this important and complex topic, this chapter includes a brief look at the nature of mathematical problem solving and an identification of the key factors that influence problem-solving performance. I have also given some suggestions for teaching problem solving. Each of these areas is organized around a few fundamental questions related to the topic of the section.

THE NATURE OF MATHEMATICAL PROBLEM SOLVING

What Is a Mathematics Problem?

Henderson and Pingry (1953) have provided an excellent set of criteria for what they call a "problem for a particular individual." These criteria are:

1. The individual has a clearly defined goal of which he [sic] is consciously aware and whose attainment he desires.
2. Blocking of the path toward the goal occurs, and the individual's fixed patterns of behavior or habitual responses are not sufficient for removing the block.
3. Deliberation takes place. The individual becomes aware of the problem, defines it more or less clearly, identifies various possible hypotheses (solutions), and tests them for feasibility. (p. 230)

Notice that whether or not a situation is a problem is determined by the individual's reactions to it. The individual must be *aware* of the existence of a situation that needs a solution and must have an *interest* in finding a solution. In addition, a procedure for determining the solution must not be immediately available. Finally, the individual must reflect on the problem in order to develop a clear understanding of what the problem is about and how to go about finding a solution. If any one of these three criteria is missing in a situation, it is *not* a problem for that individual. Thus, in order for a situation to be a problem for an individual, the person must: (1) be *aware* of the situation, (2) be *interested* in resolving the situation, (3) be *unable to proceed directly* to a solution, and (4) make a *deliberate attempt* to find a solution. A *mathematics problem*, then, is simply a problem for which the solution involves the use of mathematical skills, concepts, or processes.

Can You Illustrate How a Mathematics Situation Might Be a Problem for One Person but Not for Another?

Consider the following "potential" problems:

1. Determine the length of the hypotenuse of a right triangle whose legs have lengths of 3 units and 4 units.
2. $34575 \div 15 =$

3. If there were 12 people at a party and each person shook hands with everyone else once, how many handshakes would there be?
4. A tennis club was planning a tournament for its 12 members. Each member was to play every other member. How many matches did they need to schedule?
5. Which is the greater number $\sqrt{10} + \sqrt{17}$ or $\sqrt{53}$?

The first problem exemplifies the fact that a situation may pose different levels of difficulty depending on the problem solver's mathematical sophistication. For example, an individual who does not know what a right triangle is would find it quite difficult to get a correct answer. A person who knows what a right triangle is but who is not familiar with the Pythagorean Theorem might also be hard pressed to obtain a correct solution. Knowledge of the Pythagorean Theorem reduces the solution to simply solving the equation $x^2 = 3^2 + 4^2$ for x. Another more sophisticated solution would involve recognizing that since *3, 4, 5* is a Pythagorean triple, the hypotenuse must be 5 units. In this case there is no need to solve an equation.

The second "potential" problem appears to be a computation exercise involving division. Clearly, this is not a problem for a person who is proficient in using a long division algorithm or who has a calculator. However, this exercise could be a very challenging problem for a person who had neither understanding of an algorithm nor access to a calculator. It should be pointed out that when exercises of this sort appear in mathematics textbooks, their purpose is not to provide problem-solving experience for students but rather to provide students with practice in using standard mathematical procedures (for example, computational algorithms, algebraic manipulations, and use of formulas).

Potential problems 3 and 4 are very much alike; they are essentially the same problem in different guises. They have the same answer (66) and the solution strategy that works for one works for the other. The only real difference is that the statements *look* different. This seemingly minor distinction is often a major determinant of success in problem solving. An individual may prefer problem 3 to problem 4 merely because its setting is a party. Also, one individual may see the similarity between the two problems and consequently solve problem 4 very quickly, whereas a different individual may struggle with problem 4 even after solving problem 3 successfully.

The fifth problem illustrates a very important criterion for a situation to be regarded as a problem; namely, the individual must be *inter-*

ested in solving it. A common attitude toward a problem like this one is indifference. A reaction like "Who cares which is greater?" is typical.

What Types of Mental Activity Are Involved in the Problem-Solving Process?

Briefly, the process of problem solving involves the coordination of knowledge, previous experience, intuition, and various analytical and visual abilities in an effort to determine a workable outcome to a situation regarded as a problem; furthermore a procedure for completely determining the outcome is not already known. Of course, this does not describe what goes on when a person is actively engaged in solving a problem. One of the most perplexing aspects of problem solving is that two individuals can obtain the same solution to a problem using apparently different, but perfectly legitimate, methods. This characteristic of problem solving makes it difficult to decide on the best procedures useful in teaching problem solving.

Perhaps the most lucid thinking about mathematical problem solving has been done by the mathematician George Polya (1957, 1962, 1965). For Polya, there are four phases in the solution process: (1) understanding the problem, (2) devising a plan, (3) carrying out the plan, and (4) looking back. Polya's model of problem solving suggests that there are four distinct phases through which the successful problem solver goes. This model is valuable as a guide in organizing instruction, but it is not much help in specifying the mental processes involved in successful problem solving.

Models that attempt to explain problem-solving behavior in terms of cognitive processes have been developed by information-processing theorists, most notably by Newell and Simon (1972). Unfortunately, the efforts of Newell and Simon and their colleagues have focused primarily on "puzzle" problems (for example, Tower of Hanoi[1]), problems related to playing chess, and algorithmic problems. These types of problems do not represent the majority of problems that confront mathematics students. The works of Newell and Simon and of Polya, however, led me to attempt developing a model for solving

1. The Tower of Hanoi problem is a very commonly used task in recent psychological problem-solving research. Briefly, the problem consists of three or more circular disks of varying sizes, each having a hole in its center, and a board containing three vertically positioned pegs. The disks are initially placed on one of the three pegs in order of size from largest to smallest. The task for the individual is to transfer the disks from that disk to a particular one of the other two pegs following certain rules.

mathematics problems that points out the factors which most influence success (Lester 1978). The result of my efforts was an unrefined model containing six distinct but interrelated stages: (1) problem awareness, (2) problem comprehension, (3) goal analysis, (4) plan development, (5) plan implementation, and (6) procedures and solution evaluation. It should be pointed out that these stages do not necessarily occur sequentially during problem solving.

Stage 1: Problem Awareness. A situation is posed for the student. Before this situation is regarded as a problem, the student must realize that a difficulty exists in the sense that the student recognizes that the situation cannot be readily resolved. This recognition often follows from initial failure to attain a goal. A second component of the awareness stage is the students' willingness to try solving the problem. If the student either does not recognize a difficulty or is not willing to try solving the problem, it is meaningless to proceed.

Stage 2: Problem Comprehension. Once the student is aware of the problem situation and declares a willingness to eliminate it as a problem, the student begins the task of making sense out of the problem. This stage involves at least two substages: translation and internalization. Translation involves interpreting the information provided by the problem into terms that have meaning for the student. Internalization requires the problem solver to sort out the relevant information and determine how this information interrelates. Most importantly, this stage results in the formation of an internal representation of the problem within the individual. This representation may not be accurate at first, but it does furnish a means of establishing goals or priorities for working on the problem. It is here that the nonsequential nature of the model shows up for the first time. The accuracy of the problem solver's internal representation may increase as progress is made toward a solution. Thus, the degree of problem comprehension will be a factor in several stages of the solution process.

Stage 3: Goal Analysis. It seems that the problem solver can jump back and forth from this stage to another. For some problems it is appropriate to establish subgoals; for others, subgoals are not needed. The identification and subsequent attainment of a subgoal often aids both problem comprehension and procedure development.

Goal analysis can be viewed as an attempt to reformulate the problem so that familiar strategies and techniques can be used. It can also involve identification of the component parts of a problem. It is a process that moves from the goal itself backwards in order to separate out the different components of the problem. Thus, goal analysis actually includes more than a simple specification of given information, specification of the interrelationship of the information, and specification of the operations that may be needed.

Stage 4: Plan Development. It is during this stage that the problem solver gives conscious attention to devising a plan of attack. Developing a plan involves much more than identifying potential strategies (for example, pattern finding and solving a simpler related problem). It also includes ordering subgoals and specifying the operations that could be used. It is perhaps this stage more than any other that causes difficulty for students, and it is common to hear mathematics students proclaim after watching their teacher work a problem: "How did he ever think of that? I never would have thought of that trick." The main source of difficulty in learning how to formulate a plan of attack originates from the fact that students are prone to give up if a task cannot be done easily. But, of course, if problems can be done too easily, they are not really problems. A good problem causes initial failure, which too often results in a refusal to continue. It may also be true that students are unable to devise good plans because they have few plans at their disposal.

Another source of difficulty for students at this stage is in ordering subgoals and specifying the operations to be used. For many students, the hardest part of problem solving is in knowing what to do first and in organizing their ideas. Consequently, in addition to teaching students strategies, they must be helped to organize their thinking and planning.

Stage 5: Plan Implementation. At this stage the problem solver tries out a plan that has been devised. The possibility of executive errors arising can confound the situation at this stage. The student who correctly decides to make a table and look for a pattern may fail to see the pattern due to a simple computation error. Errors of this type probably cannot be eliminated, but they can be reduced if instruction on implementing a plan also emphasizes the importance of evaluating the plan while it is being tried. Thus, while stages 5 and 6 are distinct, they are interrelated. The main pitfalls of stage 5 are that the problem solver may forget the plan, become confused as the plan is carried out, or be unable to fit the various parts of the plan together. Fitting the parts of a plan together can be a very difficult task in itself due to the fact that the best sequencing of steps in the plan or the best ordering of subgoals may not be clear to the problem solver. For some problems, the sequencing of subgoals does not matter, while for others, it is essential that subgoals be achieved in a particular order.

Stage 6: Procedures and Solution Evaluation. Successful problem solving is usually the result of a systematic evaluation of the appropriateness of the decisions made during the problem-solving process and a thoughtful examination of the results obtained. The role of evaluation goes far beyond just checking the answer to be sure that it makes sense. Instead it is an ongoing process that begins as soon as the problem solver begins goal analysis and continues long after a solution has been found. Procedures and solution evaluation may be viewed as a process of seeking answers to certain questions as the problem solver works on a problem. Representative of the questions that should be asked at each stage are the following:

1. Problem comprehension—the problem solver evaluates how well he or she understands what the problem is: (a) What are the relevant and irrelevant data involved in the problem? (b) Do I (the problem solver) understand the relationships among the information given? (c) Do I understand the meaning of all the terms that are involved?
2. Goal analysis—the problem solver categorizes the information into classes like givens, operations, variables, and so forth and attempts to identify the structure of the problem: (a) Are there any subgoals that could help me achieve the goal? (b) Can these subgoals be ordered? (c) Is my ordering of subgoals correct? (d) Have I correctly identified the conditions operating in the problem?
3. Plan development—the problem solver searches for a method of proceeding: (a) Is there more than one way to do this problem? (b) Is there a best way? (c) Have I ever solved a problem like this one before? (d) Will the plan lead to the goal or a subgoal?
4. Plan implementation—the problem solver tries out a plan: (a) Am I using this strategy correctly? (b) Is the ordering of the steps in my plan appropriate or could I have used a different ordering?
5. Solution evaluation—the problem solver analyzes the results: (a) Is my solution generalizable? (b) Does my solution satisfy all the conditions of the problem? (c) What have I learned that will help me solve other problems?

Figure 2-1 attempts to illustrate the interrelationships that exist among the stages in the model. It also suggests how a student might proceed in solving a problem. Finally, it can be useful to the teacher

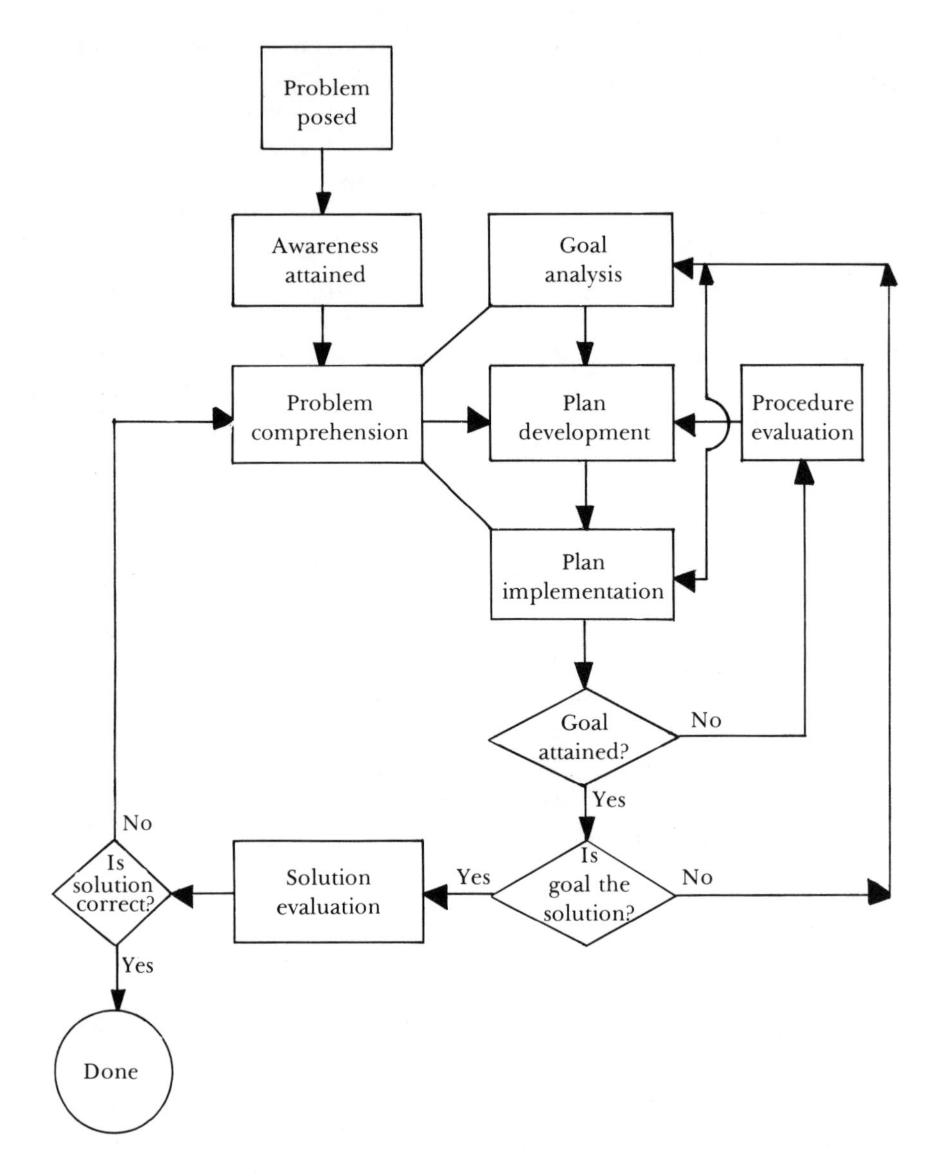

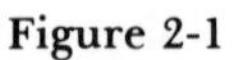

Figure 2-1

Schematic representation of a model of mathematical problem solving

who is trying to organize appropriate problem-solving experiences for students by highlighting various potential sources of difficulty.

FACTORS THAT INFLUENCE SUCCESS IN PROBLEM SOLVING

What Are the Main Factors That Affect Success In Problem Solving?

The model may help identify points during problem solving at which an individual could have difficulty, but it does not describe the factors that make a problem difficult or cause one problem solver to succeed while another has difficulty. A wide range of factors influence success in problem solving. These factors fall into four distinct but interrelated categories: the problem, the problem solver, the problem-solving processes, and the problem-solving environment.

Category 1: The Problem. The nature of the problem itself affects success. The mathematical content of a problem is certainly important as are the format of the problem (that is, the way the problem is posed or represented) and the structure of the problem (for example, linguistic, syntactic, or logical structures).

Category 2: The Problem Solver. Various characteristics of the individual play an important role in problem solving. These include: mathematical background and sophistication, prior experience with similar problems, reading ability, conceptual style of thinking, ability to persevere, tolerance for ambiguity, resistance to premature closure, and reaction under stress. Of course, this list is not exhaustive but merely representative.

Category 3: The Problem-Solving Process. Process factors are very closely linked with both problem factors and problem-solver factors. This category involves factors related to the behavior of the individual during problem solving. The manner in which the problem solver organizes and processes information, the types of cognitive strategies used to plan and carry out a plan, and the methods used to evaluate what was done comprise this category. Key factors in this category that influence success are the complexity of acceptable solution processes, the problem solver's familiarity with useful solution processes, and demands made on idea getting and idea evaluating.

Category 4: The Problem-Solving Environment. Features of the problem-solving environment that are external to the problem and the problem solver are included in this category. External forces such as instruction, the conditions under which the individual must work, and time constraints are representative of the factors in this category.

Of course the factors associated with these four categories interact with each other throughout the course of solving a problem, and the interactive nature of these factors is what causes problem solving to be such a complex type of human activity. This complexity can be demonstrated by considering several of the factors known to play a large role in determining the difficulty of a problem. In the following section, seven factors are listed and discussed. These seven factors represent a composite of the ideas of several persons who have devoted considerable attention to the nature of problem solving as well as ideas I have gleaned from my own personal problem-solving activities and from observing many students working on problems. It is noteworthy that each of these seven factors falls into at least two of the four categories just discussed. These seven factors provide at least a partial answer to the following question.

What Makes a Problem Difficult?

The factors contributing to the difficulty of a problem are probably too numerous for an exhaustive listing of them all. However, there does seem to be a reasonably small number of factors that account for much of the difficulty an individual can have with a particular problem.

Complexity of the Problem Statement. The amount of information, number of variables, syntactical complexity, and the mathematical content of the problem are among the factors which greatly influence problem difficulty (categories 1, 2).

Method of Problem Presentation and Representation. Success in solving a problem is greatly influenced by the way the problem is posed for the problem solver and the context within which it is represented (categories 1, 2, 4).

Familiarity of the Problem Solver with Acceptable Solution Procedures. A problem whose solution calls for the use of a particular solution is easier if the problem solver has used that procedure before (categories 2, 3).

Misleading Incorrect Solutions or Solution Procedures. Oftentimes a problem suggests a solution or procedure that is incorrect or ultimately of no help (categories 1, 3).

Difficulty in Locating Reachable Subgoals. The complaints "I don't know how to get started" or "I don't know what to do first" are common among unsuccessful problem solvers (categories 1, 2, 3).

Constraints Arising from Misconceptions or Misunderstanding of Information Given in the Problem. Misunderstanding or overlooking information in a problem statement is a very common cause of failure. Also, problem solvers sometimes fail to get a correct solution because they impose a constraint that leads them to work only within the framework of the constraint (categories 1, 2).

Affective Factors Associated with the Problem Solver's Reaction to the Problem. Lack of motivation and perseverance, high degree of stress, and low tolerance for ambiguity are illustrative of these factors. Problems of this type are legion and are linked with the individual. Some problems are too "messy" to bother with, while others are not interesting to some individuals because they do not enjoy that kind of problem (categories 1, 2, 3, 4).

INSTRUCTION IN PROBLEM SOLVING

The teacher who gives conscious attention to helping students overcome the obstacles presented above will be taking some very positive steps toward improving their problem-solving performance. Unfortunately, instruction that is restricted to helping students avoid error or difficulty is inadequate. What is needed is instruction that is concerned not only with sources of difficulty but also with promoting the development of good problem-solving habits (for example, self-motivation, perseverance, resistance to premature closure) and with the acquisition by students of a wide range of skills and strategies (for example, making tables to organize data, estimating, looking for patterns, working backwards, and solving a simpler problem). Indeed, even with the knowledge that there are four categories of factors that influence problem-solving success and that seven factors are especially prominent among all the factors that make a problem difficult, an important question arises.

How Should Students Be Taught to Be Better Problem Solvers?

Unfortunately, the identification of the *best* ways to teach problem solving is not easy. It is hard to know exactly what causes changes in problem-solving behavior. The four categories of factors discussed in the preceding section are inextricably linked together in a way that makes it seem practically impossible to know what sort of instruction maximizes improvement in problem-solving performance.

A number of differing viewpoints regarding instruction in problem solving have been proposed. The most common of these are based on the very thoughtful writings of Polya, whose four-stage model, described earlier, has direct applicability to instruction. Attempts to develop instructional methods using Polya's ideas typically focus on teaching students various heuristic strategies; that is, planned actions or series of actions performed to assist in the discovery of a solution to a problem. There are other instructional methods in addition to those patterned after Polya's suggestions. These include: (1) have students solve many problems without specific intervention by the teacher; (2) teach certain unitary tool skills (for example, making tables, drawing diagrams, and translating from written form to equation form); (3) model good problem-solving behavior and have students imitate this behavior; (4) teach heuristic strategies; (5) some combination of the preceding. A brief discussion of these methods may help give a clearer view of what each involves.

A teacher using method 1 would select a large number of problems on the basis of certain criteria. For example, a fifth-grade teacher who wants students to develop some facility in solving process problems might use the following criteria: mathematical content must be no higher than fifth-grade level; problems must be interesting to the students; solution processes must be within the grasp of fifth graders; some problems should have more than one answer (others no answer); there should be more than one way to solve each problem; and some problems should be related to others in the sense that they are similar to, build on, or are extensions of other problems. A problem bank would be placed in a mathematics corner, and students would be encouraged to work several problems during a given period of time. From time to time the teacher would lead a class discussion about attempts to solve certain problems. Beyond summarizing discussion, the teacher would play no role other than classroom manager.

Instruction to teach particular "tool skills" useful in problem solving is common in many current mathematics textbooks. Tool skills are skills that facilitate planning an attack, help in organizing relevant information, and otherwise assist the problem solver in using a strategy. In general, the tool-skills approach attempts to teach a few carefully chosen "facilitating" techniques, such as making a table, writing an equation, using a formula, or making estimates. For example, instruction in making a table might begin with a demonstration by the teacher of how to make a table to solve a specific problem. The teacher would point out how the table helps to organize and keep track of information. Subsequently, students practice reading and constructing tables. Finally, they are asked to solve several problems for which solutions are made easier by making tables.

The third instructional method, modeling good problem solving, is used at every level—elementary school through graduate school. This method has been characterized as the method of "teaching *about* problem solving" (Hatfield 1978). The teacher demonstrates how to solve a certain problem and directs the students' attention to salient procedures and strategies that enhance the solution of the problem. Students are then expected to solve problems using the processes modeled by the teacher.

The method of instruction in heuristics has been by far the most popular one studied in recent years by researchers in mathematics education. This popularity probably stems from a general acceptance of Polya's four stages in problem solving and the questions associated with each stage (Polya 1957). The heuristic method involves the presentation of the content of mathematics in the context of problems to be solved. The reader who is interested in learning more about teaching heuristic strategies is referred to books by Polya (1957, 1962, 1965) and Rubinstein (1975) and the article by LeBlanc (1977).

In my opinion, some combination of these four methods is probably the most sensible approach to take. Certainly, students will not improve their problem-solving skills unless they try to solve a wide range of types of problems. It also seems obvious that the likelihood of improved problem-solving performance is increased if students see good problem-solving behavior exhibited by their teacher. It is at least safe to say that the inverse of this statement is probably true. That is, if the teacher does *not* exhibit good problem-solving behavior, it is unlikely that the students' ability to solve problems will increase. Finally demonstration by the teacher of the usefulness of various tool skills and heuristics can clearly do students no harm, and it is probable that such demonstrations will enhance good problem solving. In short, while I cannot recommend any one method over the others, conscious attention to any or all of these methods will have a positive effect on the problem-solving performance of students.

Are There Any Guidelines to Follow in Planning Problem-Solving Activities for Students?

Any list of guidelines is fraught with shortcomings because of the many variables involved in problem solving. Just as there is no stock set of guidelines for good teaching, there is no fixed set of guidelines for teaching problem solving that applies to all situations. Nonetheless, several suggestions merit serious consideration by every mathematics teacher. These suggestions were proposed by Brownell (1942) nearly forty years ago. His ideas about mathematical problem solving are just as timely and provocative today as when he first proposed them. His list of "practical suggestions for developing ability in problem solving" is given below in a somewhat abbreviated form.

1. Whether a learning task is a problem to the learner depends upon peculiar relationships between the learner and the task.
2. There is little that is educational in attempts to solve puzzles as puzzles. When a puzzle must be presented, the learner can be helped to analyze the task and to organize his attack, the puzzle thus being converted as much as possible into a problem.
3. When a problem is posed, the relationship necessary to its solution should be well within the understanding of each child and identifiable by him or her with reasonable effort.
4. Teaching should start with whatever technique the child uses proficiently and should guide the child in the adoption and use of steadily more mature types of problem solving.
5. Meanings and understandings are most useful in problem solving when they have themselves been acquired through solving problems.
6. To the limits desirable and possible, solutions to problems should be summarized clearly, stated verbally, and generalized.
7. Practice in problem solving should not consist in repeated experiences in solving the same problems with the same techniques but should consist in the solution of different problems by the same techniques and in the application of different techniques to the same problems.
8. A problem is not truly solved unless the learner understands what he or she has done and knows why his or her actions were appropriate.
9. The mistakes students make when they are trying to solve problems are not "corrected" by providing them with the "right" solution. They are correctly solved only when (a) the weakness in technique has been exposed and supplanted by a sounder attack or (b) the needed meaning or understanding has been developed or (c) when both (a) and (b) have been taken care of.

10. Instead of being "protected" from error, the child should be exposed to error and be encouraged to detect and demonstrate what is wrong and why.
11. An inquiring and questioning attitude toward problem solving is produced best by continued experience in solving real problems, one consequence of which is that the learner comes to *expect* new problems and to look for them.
12. Guidance of learning to solve a particular problem may well take the following form: (a) help students formulate the problem clearly; (b) see that they keep the problem continuously in mind; (c) encourage them to make many suggestions by having them analyze the situation, recall similar cases and the rules or principles of solution, and guess courageously; (d) have them organize their solution process by building outlines, using diagrams and graphs, taking stock from time to time, and formulating concise statements of the net outcomes of their activity.

There is one very important element missing from Brownell's list of suggestions, although it is implied. This element involves the attitude the teacher has toward problem solving and it can be summed up in the following way: Problem-solving instruction is most effective when students sense two things: (1) that the teacher regards problem solving as an important activity, and (2) that the teacher actively engages in solving problems as a regular part of mathematics instruction.

Thus, it is not enough for the teacher to collect a large number of good problems and to follow the twelve guidelines proposed by Brownell. Teachers who want to maximize the chances that their students will improve their problem-solving abilities must make students believe in the value and importance of problem solving as a part of doing mathematics and must themselves be problem solvers. The maxim "Do as I say, not as I do" will have no credibility with students as far as problem solving is concerned.

FINAL COMMENTS

I have tried to convey the message that problem solving is the most subtle and complex of all types of learning and that it is also the most important goal of doing mathematics. I have pointed out the extreme complexity of problem solving by noting that it is a very personal kind of intellectual activity which is influenced by a wide variety of interrelated factors. Despite all of this, there are some things teachers can do to enhance the development of problem-solving skills and processes in students. With these ideas in mind a final question arises.

Should Students Be Taught to Be Better Problem Solvers?

I believe strongly that a thoughtful teacher can have a very positive effect on students' problem-solving behavior. But it is not easy. Good instruction in problem solving requires attention by the teacher to a myriad of factors that affect success. At the same time, it necessitates frequent, continuous, and active participation by students. Using a somewhat forced analogy, perhaps I can illustrate why it is necessary for a teacher to give direct attention to teaching problem solving. Learning how to solve problems is like learning how to play baseball. Just as one cannot expect to become a good baseball player if one never plays baseball, a student cannot expect to become a good problem solver without trying to solve problems. Also, the overwhelming majority of baseball players develop their abilities under the guidance of thoughtful coaches who keep them from developing bad habits and point out the most appropriate ways to throw, slide, bunt, and so on. Likewise, most students will fully develop their problem-solving abilities only if they are kept from acquiring bad habits and are made aware of various techniques and processes that are useful in problem solving. Of course, the analogy is flawed by the fact that many children begin playing baseball long before they receive any coaching, whereas relatively few students are self-motivated solvers of problems in mathematics. It is precisely this weakness in the analogy that makes the argument for instruction in problem solving compelling. That is, very few students develop their problem-solving abilities to their full potential *on their own*. Thus, since "learning to solve problems is the principal reason for studying mathematics," it is vital that school mathematics instruction give conscious attention to problem solving. It is my belief that the ideas and guidelines discussed in this chapter will help teachers to develop perspective about the nature of mathematical problem solving, which in turn will aid them in planning appropriate problem-solving experiences for their students.

REFERENCES

Brownell, William A. "Problem Solving." In *The Psychology of Learning*. Forty-first Yearbook of the National Society for the Study of Education, part 2, edited by Nelson B. Henry. Chicago: University of Chicago Press, 1942, pp. 438–40.

Hatfield, Larry L. "Heuristical Emphases in the Instruction of Mathematical Problem Solving: Rationales and Research." In *Mathematical Problem Solving: Papers from a Research Workshop*, edited by Larry L. Hatfield and David A. Bradbard. Columbus, Ohio: ERIC/SMEAC, 1978.

Henderson, Kenneth B., and Pingry, Robert E. "Problem Solving in Mathematics." In *The Learning of Mathematics: Its Theory and Practice*. Twenty-first Yearbook of the National Council of Teachers of Mathematics. Washington, D.C.: The Council, 1953, pp. 228–70.

LeBlanc, John F. "You Can Teach Problem Solving." *Arithmetic Teacher* 24 (November 1977): 16–20.

Lester, Frank K. "Mathematical Problem Solving in the Elementary School: Some Educational and Psychological Considerations." In *Mathematical Problem Solving: Papers from a Research Workshop*, edited by Larry L. Hatfield and David A. Bradbard. Columbus, Ohio: ERIC/SMEAC, 1978.

National Council of Supervisors of Mathematics. "Position Paper on Basic Mathematical Skills." *Arithmetic Teacher* 25 (November 1977): 19–22; also in *Mathematics Teacher* 71 (February 1978): 147–52.

Newell, Alan, and Simon, Herbert A. *Human Problem Solving*. Englewood Cliffs, N.J.: Prentice-Hall, 1972.

Polya, George. *How to Solve It*. 3d ed. Garden City, N.J.: Doubleday, 1957.

Polya, George. *Mathematical Discovery: On Understanding, Learning, and Teaching Problem Solving*. Vol. 1. New York: Wiley, 1962.

Polya, George. *Mathematical Discovery: On Understanding, Learning, and Teaching Problem Solving*. Vol. 2. New York: Wiley, 1965.

Rubinstein, Moshe F. *Patterns of Problem Solving*. Englewood Cliffs, N.J.: Prentice-Hall, 1975.

3. The Role of Computation

Thomas E. Rowan

During March of 1978 a National Conference on Achievement Testing and Basic Skills was held in Washington, D.C. In his message to that conference President Carter stated, "There is no greater challenge than the one facing our educational system . . . to ensure that all children learn, at the very least, to read, to write, and to compute" (U.S. Dept. HEW 1979). This statement by the president of the United States captures the central issue of this chapter in a few words, "to ensure that all children learn, at the very least . . . to compute." There is no doubt that the general public currently views this as the fundamental goal of mathematics education, and if we do not succeed in this regard, then we do not succeed at all, according to the view of many educational observers.

In contrast, the National Advisory Committee on Mathematics Education (NACOME), Conference Board of the Mathematical Sciences in their report *Overview and Analysis of School Mathematics, Grades K–12* (1975) states:

> . . . it appears to us that the case for decreased classroom emphasis on manipulative skills is stronger now than ever before. Impending universal availability of calculating equipment suggests emphasis on approximation, orders of magnitude, and interpretation of numerical data . . . not drill for speedy, accurate application of operational algorithms. (p. 24)

To some extent at least, these two statements can be interpreted as expressing differing extremes on the central topic of this chapter, computation. Is the ensurance of computational skill the "great challenge" of our educational system, or is it an outdated goal that needs to be deemphasized in the face of a changing world? Views on the importance of computation, from a relative as well as an absolute sense, vary widely both among and between mathematics educators and clients of the educational system. The purpose of this chapter is to discuss the current status and to suggest the possible future of computation in mathematics education.

CURRENT STATUS OF COMPUTATION

Computation skill is one dimension of both elementary and secondary mathematics education that would probably be considered of prime importance. Much of this importance derives from the general public and the popular press, and to some extent, this is a reaction to reports of declining scores on standardized tests. Much discussion (Munday 1979) has centered around whether the decline in test scores is real or a result of the characteristics of the testing instruments, testing methods, or other factors. A very real part of the debate has been about the extent to which the curricular reforms of the late 1950s and early 1960s contributed to the suspected decline. Adequate data do not currently exist to respond clearly to the main issue of whether there is a real decline and so it is certainly not possible to intelligently analyze the secondary issue of causal factors.

The situation with respect to test data was summarized in the NACOME report:

> . . . there has been a tendency for the traditional classes to perform better on computation while modern classes perform better in comprehension. There appears to have been a decline in scholastic skills since 1960. Mathematics achievement has shared in that decline. . . . In general, data that would support a definitive picture of mathematics in grades K–12 do not exist. The national picture is far more varied and complex than either proponents or critics of recent curricular innovation suggest in their public debates. (pp. 118–19)

Whether or not the test scores accurately reflect a decline in computational proficiency does not change the effect of the publicity they received. Recent years have seen the back-to-basics slogan emerge. Most states have taken action to initiate testing programs that measure the acquisition of specific skills, usually labeled "minimal competencies," including computational skills. In many cases a student must achieve a predetermined score on the test to be eligible for high school graduation.

The *New York Times* of April 22, 1979, carried a "spring survey of education." An article included in that feature was entitled, "Beyond the New Math." Nancy Rubin, a freelance writer stated:

> Increasing accountability pressures by states and local school districts and a widespread "back-to-basics" movement has propelled mathematics education back into the more familiar—if not more shallow—waters of computation achievement.

Very few writers in the popular press have taken this kind of position with regard to the movement toward the basics, and computation is the familiar ingredient of the mathematics program for most people. It is understandable that they react with concern when test scores seem to be declining, and there is a generally held belief that computational skills have been deemphasized in the school curriculum.

As the public reacted to circumstances it perceived as weaknesses in the school mathematics program, the mathematics education community has in turn reacted to the back-to-basics trend. In 1977 the National Council of Supervisors of Mathematics (NCSM) published a position paper on basic skills (NCSM 1978). This paper represented the thoughts of many mathematics educators who were responding to questions and expressing their ideas on basics. Later this position paper was published in both major journals of the National Council of Teachers of Mathematics (NCTM) for the consumption of elementary and secondary schoolteachers. The position taken is that the "basics" include problem solving, applications, geometry, measurement, and charts, tables, and graphs. It also reflected the position taken by NACOME (cited in the introduction to this chapter) by including reasonableness of results, approximation, prediction, and computer literacy as basic skills. However, NCSM did not call for a deemphasis of computation. They simply asked that it be "put into its proper place." The position paper stated:

> Any list of basic skills must include computation. However, the role of computational skills in mathematics must be seen in the light of the contributions they make to one's ability to use mathematics in everyday living. In isolation, computational skills contribute little to one's ability to participate in mainstream society. Combined effectively with the other skill areas, they provide the learner with the basic mathematical ability needed by adults. (p. 148)

The NCSM position paper attempted to put computational skills into a perspective that would be acceptable to both its proponents and detractors. It remains to be seen whether this will successfully stop what has sometimes been referred to as the pendulum characteristic of educational change.

Another aspect of the current status of computation deserves mention. Looking again at the NACOME report, we find these statements:

> Conceptual thought in mathematics must build on a base of factual knowledge and skills. But traditional school instruction far overemphasized the facts and skills and far too frequently tried to teach them by methods stressing rote memory and drill. These methods contribute nothing to a confused child's understanding, retention, or ability to apply specific mathematical knowledge. Furthermore, such instruction has a stultifying effect on student interest in mathematics, in school, and in learning itself.
>
> The members of NACOME view with dismay the great portion of children's school lives spent in pursuing a working facility in the fundamental arithmetic operations. For those who have been unsuccessful in acquiring functional levels of arithmetic computation by the end of eighth grade, pursuing these skills as a sine qua non through further programs seems neither productive nor humane. We feel that providing such students with electronic calculators to meet their arithmetic needs and allowing them to proceed to other mathematical experience in appropriately designed curricula is the best policy. (pp. 24, 25)

Three issues are clearly set out in these statements. The two that have received some attention both outside and inside the mathematics education community are the definition of appropriate minimal competencies and the role calculators should play in mathematics learning. The third issue, which is better known within than outside the mathematics education community, is that of teaching-learning practices. Have we overemphasized facts and skills at the expense of understanding, retention, and ability to apply mathematical knowledge? If we have, what can be done about it and how can this change be accomplished without starting another pendulum swing?

COMPUTATION IN THE 1980S

Predicting the future of computation is risky at best. The most accurate prediction would probably be to say simply that there will be very little change; anyone who works in education is certainly aware that changes usually occur very slowly. Without taking the "safe" stance, however, the purpose of this chapter is to explore what might occur and why. Many of the projections may not occur, but hopefully the direction of the current will be reasonably approximated by these speculations. An effort will be made to include each of the dimensions discussed in the section on status.

The NCSM position paper on basic skills offers a compromise between those who feel computation should be deemphasized and those who feel it should be the primary purpose of school mathematics. It is a consideration worth attending to as we come to grips with the problem of defining basic skills. The definition that evolves and becomes generally accepted will have far-reaching implications for the issues of testing and minimal competencies as well as for teaching and learning practices.

Those in the mathematics community who advocate a direct deemphasis of computational skills would do well to reassess this posture. Suggesting the possibility of such a radical change is enough to cause many people in the general public to sever all consideration of a well-rounded mathematics program. It would be better to keep the communication lines open by accepting computational skills as a valid part of the mathematics program and then going on to discuss other important parts that should not be neglected.

In addition to its importance to the general public, and perhaps to some extent underlying that importance, there are other reasons for predicting that computation will retain a central role in the mathematics program. Computation is both a tool as well as a key to open doors for other aspects of mathematics. For example, one needs computation to check the reasonableness of results obtained from work done on a calculator or computer. Measurement is certainly a basic skill, but computation is needed in measurement—to find areas, volumes, and perimeters for example. Computation also assists in the solution of problems in areas such as probability. In the case of algebra, computation is both a tool and a model from which many algebraic skills and principles can be derived. Finally, without the confidence built from properly taught computational skills, many students might not venture into the more advanced areas of mathematics.

Computation is a tool that can either facilitate or hinder leaps to higher mathematical generalizations. A popular anecdote from the history of mathematics is the story of how Karl Frederick Gauss, the distinguished mathematician of the nineteenth century, solved the busy-work problem of summing the numbers from 1 to 100. He realized that by pairing the numbers ($1 + 100$, $2 + 99$, $3 + 98$, and so forth) he would have fifty pairs, each with the sum of 101, and that the simple computation of 50 times 101 would give him the answer to the problem. Both the realization as well as the final solution owe some debt to Gauss's computational abilities.

The need for an expanded definition of basic skills that would be acceptable to the majorities in mathematics education and the general public was reinforced in the February 1980 newsletter of the National Assessment of Educational Progress (NAEP). The first paragraph of that newsletter stated:

> Calling for an expanded definition of "basic skills" in mathematics, educators testifying before a U.S. House of Representatives subcommittee said students apparently are learning the mathematics skills being taught in schools but that simply mastering those skills may not be enough.

Later, the same newsletter exlained:

> Assessment results showed "satisfactory" performance on knowledge items and computational skills, but weaker abilities on problem-solving tasks and with more complex skills such as fractions, decimals, and percents.

These NAEP statements point up both the need for good computational skills *and* the danger of overemphasizing computation or teaching it in isolation from other important components of mathematics.

Perhaps this is more a hope than a prediction: future definitions of basic skills should include, but not be limited to, computation. Bad experiences with current minimal competency tests that restrict themselves to trivial aspects of mathematics should lead to the development of better tests. It is unlikely that public pressure for some type of minimal competency test will cease during the next ten years. To the contrary, the declining economic situation and reduced employment opportunities will likely increase public concern over educational outcomes. If test developers can meet the challenge and produce instruments that sample the range of desirable outcomes, then the mathematics program will not be adversely affected by this trend.

The next several years may see considerable growth in teaching-learning practices. The work of Piaget is just now beginning to have significant impact on curriculum and classroom practices in the United States. More recently, studies related to left and right brain function have produced interesting implications for teaching and learning. Evidence should begin to accumulate that curricula and practices based on these developments are beneficial in an observable manner when studied longitudinally, and this should lead to more efficiency and better attitudes in computational learning. If this occurs, it will make computation more acceptable to mathematical educators who currently feel much time is wasted on it. At the same time it will satisfy public demands for improved computational skills while providing increased time for the study of other important areas, such as probability and geometry.

Minor adjustments to the curriculum itself will also occur. Computation with common fractions will become less important as the metric system gradually gains ground. Such computation will be delayed until later grade levels and decimal-fraction computation will come earlier in the curriculum, probably beginning with concrete models and extended place value in grade two. Such curricular changes will be relatively minor when compared to the likely changes in teaching-learning practices.

Areas in which extensive change is almost certain are those that are influenced by changes in calculating and computing equipment. Calculators have now become so common in the American household that resistance to their appropriate use in mathematics classrooms should rapidly disappear. People who use calculators in their daily lives will know full well that such a device does not replace thinking. The fact that the equipment does *not* solve the problem will be clear. The demonstration of the beneficial effects of calculators on computational skills, while they simultaneously open up new areas of study, will increase their acceptance by teachers and citizens alike.

The most recent technological innovation is the microcomputer. The effect that it will have on computational skills may be even more dramatic than that achieved by the most effective use of calculators. As microcomputers become inexpensive, readily available, and reliable, they will be used increasingly to provide individualized practice and/or instruction for students who have difficulty with computation. They can diagnose, at least to the initial levels of algorithm development. They can maintain student records and provide teachers with information to improve instructional decisions. They motivate students and permit them to exercise their creativity in analyzing and solving problems that can be attacked through computer programs. They are already simple enough so that their use requires virtually no special training, and students can use them with a minimum of supervision. In short, they are an innovation that could be applied almost overnight. As better software is developed, they will become increasingly attractive to schools.

The combination of improved teaching-learning practices at the earlier levels and individualized attention through microcomputer technology may well have a positive effect on computational skills that exceeds anything we can currently imagine. Of course, the effect is predicated upon the implementation of both factors.

In summary, the future of computation appears to be a promising one. Computation will probably not be deemphasized, but at the same time it will probably not dominate at the expense of other mathematical goals. Computation will probably be taught more efficiently because of advances in what we know about learning and because of new and exciting technology.

REFERENCES

Munday, Leo A. "Changing Test Scores, Especially Since 1970." *Phi Delta Kappan* 60 (March 1979): 496–99.

National Advisory Committee on Mathematical Education (NACOME). *Overview and Analysis of School Mathematics.* Washington, D.C.: Conference Board of Mathematical Sciences, 1975.

National Assessment of Educational Progress (NAEP). *NAEP Newsletter*, vol. XIII, no. 1 (February 1980).

National Council of Supervisors of Mathematics (NCSM). "Position Paper on Basic Mathematical Skills." *Mathematics Teacher* 71 (February 1978): 147–52.

U.S. Department of Health, Education, and Welfare. *The National Conference on Achievement Testing and Basic Skills.* Washington, D.C.: Department of Health, Education, and Welfare, 1979, p. 21.

4. Measurement: How Much?

Alan Osborne

Measurement concepts, processes, and skills belong in the comprehensive curriculum. We all recognize and appreciate the role measurement plays in the everyday life of our technological, consumer-oriented society, and no one seriously considers decreasing the attention given to measurement in the science and mathematics programs in the schools.

Performance on measurement tasks in a variety of assessments leaves something to be desired. Recently released data from the mathematics portions of the National Assessment of Educational Progress (1979b) suggest that more attention should be given to helping students acquire more extensive repertoires of understandings and skills in measurement. Of particular note is the deficient performance in the areas concerned with applications of mathematics (1979c). The apparent answer to the titular question of "How much?" is simply "More."

However, the decision to increase the time given to instruction on measurement in the school mathematics program generates a host of questions, issues, and problems. There is a finite amount of time in a school day, and most school systems have a ceiling of approximately 180 days for instruction. To increase the time allotted to instruction in measurement means either increasing the time given to instruction in mathematics or deemphasizing other topics within mathematics in order to make room for measurement. To argue for an increase in time given to measurement by decreasing time allotted other mathematical topics that are neither so fundamental nor basic, it must also be noted that several currently underrepresented topics, such as statistics and computer literacy, are exceedingly important components of any program of universal literacy in mathematics. These topics are legitimate competitors with measurement for representation in the curriculum for all students. Indeed, in making the case for a comprehensive curriculum in mathematics, it is difficult to justify allotting more time for instruction in measurement. Rather, the significant issues and problems should focus on how to make more effective and efficient use of time allotted to the topic of measurement.

Changes in instructional practices and in curricular themes that have the potential for improving the teaching of measurement are identified in the following sections. Each concerns the efficiency and effectiveness of instruction, and three specific areas are identified for change. The first concerns organizing instruction to focus on the fundamental characteristics of measurement that are at the heart of any measurement system. They form the structural foundation that enables the student to eventually transfer what was learned from one system of measurement to another. In this way, measurement learning can be made more effective and efficient. The second area of change concerns taking better advantage of measurement ideas to teach numerical skills and understandings. Most explanatory models for numerical operations are based on measurement concepts. The research evidence demonstrates that judicious use of measurement models improves performance with numbers, but we have seldom exploited our instructional strategies to focus on the contribution of these strategies to learning about measurement; measurement is only part of the implicit backdrop for number. Taking systematic advantage of the measurement context can lead to improved understanding about measurement with considerable gain in instructional efficiency. The third area of change is in the teaching of estimation strategies. Estimation is rarely taught since activities are not presented in the textbook contents. If estimation strategies are taught, however, they serve to reinforce and elaborate the fundamental structural concepts of measurement.

RESPECTING THE NATURE OF MEASUREMENT

Most instructional programs about measurement are not organized in terms of the major concepts of measurement. Learning activities address three different goals that are important in their own right, but they do not help learners form a sufficiently powerful ideation structure to enable them to deal successfully with new learning and problem solving involving measurement. First, instruction focuses on using the tools of measurement. Learning to use rulers, thermometers, measuring cups, scales, and the like is important. However, without a sense of the nature of the relationship between these tools and what is being measured, little insight into the scientific and mathematical processes is obtained. Second, instruction emphasizes the use of formulas and computational rules at too early a stage. Hirstein, Lamb, and Osborne (1978) document some misconceptions about area measure; they indicate that the probable source of the students' lack of understanding is insufficient experience with preliminary, primitive measurement concepts prior to dealing with area on a rule basis. Particularly when dealing with measurement systems for length, area, and volume, learners must have the opportunity to tie together a base of geometric understanding with the actual number concepts of measurement. Finally, teachers have assumed a responsibility for helping children acquire a familiarity with metric measurement. This is a laudable goal, and the evidence of the recent and former National Assessments (1979a) indicates that the goal is being realized. However, instruction has been directed more to establishing familiarity with and use of the metric units than in using learning about metric measurement to reveal the nature of measurement.

The quality of current instruction about measurement in most school mathematics programs is such that most youth leave school with several particulate bits of discrete knowledge about measurement rather than a feel for and an understanding about the nature of measurement as a system. If a learner acquires an understanding of the characteristics of a measurement system, then the learner possesses an ideational structure that can serve as a base both for problem solving and for measurement ideas. These are:

1. Number assignment: To measure an object is to assign a number to an attribute or a state of the object. For example, the area of a rectangle is the number of units—perhaps square centi-

meters, acres, or square feet—that it would take to cover the object.

2. Comparison: If object A is "contained by" object B, then the measure of object A is less than the measure of object B. Identified by many researchers as the most primitive of the measurement concepts, this is a good place to begin instruction with young children. Children can compare objects to see which one is bigger or smaller in a number-free sense. Later in the instructional sequence, care must be taken to incorporate number into the comparisons.
3. Congruence: If object A and object B are congruent, then the numbers that are their measures are the same.
4. Unit with iteration: There is a special object or state in any measurement system to which the number one is assigned. The object with measure one is the base for identifying the functional rules that determine the number assignment for any object or state. The iteration provides for the covering of the object that gives the child the basis for counting the measurement. The research evidence suggests that children who acquire an understanding of unit and iteration of the unit for counting have greater power in dealing with measurement both in their early learning about measurement and in later learning of more sophisticated elaborations of measurement concepts. Thus, it appears wise to have children find the length of objects by covering them with units and counting the units for an extensive period of time prior to using rulers. In the same way, in dealing with areas and volumes, early experiences should stress the covering or filling with units along with considerable counting activity before proceeding to the use of formulas.
5. Additivity: This basic structural characteristic of measurement systems is more readily exemplified than verbally defined. Figure 4-1 exhibits the additivity property in terms of measurement of length. Note that the length of the join of two segments is the same as the sum of the lengths of the two segments. The property assures that the geometry of the objects is mirrored in the operations with the number that are the measures. Children who have usable control and understanding of this concept of additivity appear to have much better and complete understanding of measurement generally.

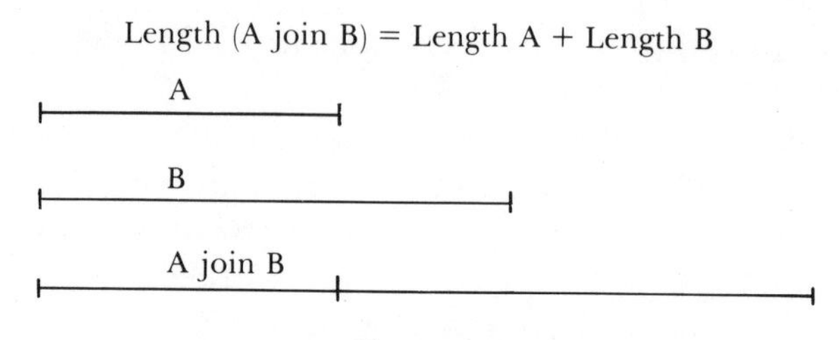

Figure 4-1
Additivity for length measurement

These five ideas serve to characterize most measurement systems. There are exceptions: temperature does not have the additivity characteristic; if you mix 40-degree cream with your 95-degree coffee, the cream cools the coffee rather than warming it to 135 degrees. But children should be led to expect every measurement system to have one or more of these characteristics. These ideas provide a basis for transfer of learning from one measurement system to another. They can be used in instruction to reduce the particulate, discrete nature of learning in measurement that typifies so many current instructional programs.

In particular, the ideas of unit, comparison, and additivity appear to be key concepts for children to grasp if they are to learn about measurement efficiently and be able to use measurement ideas. Research evidence suggests that organization of learning experiences through activities using materials and objects that are focused on these concepts prior to experiences with rulers and mensuration formulas is the most effective way to promote conceptual growth and understanding of measurement.

The 1976 Yearbook of the National Council of Teachers of Mathematics entitled *Measurement in School Mathematics* (Nelson 1976), *Number and Measurement: Papers from a Research Workshop* (Lesh 1976), and *A Metric Handbook for Teachers* (Higgins 1974) all contain papers that elaborate on the mathematical and scientific nature of measurement and examine the foundations of measurement in terms of the development of concepts in children.

USING CONCEPTS OF MEASUREMENT IN TEACHING NUMBER

Most ideas and skills are better learned if they are used. Instruction about number frequently overlooks opportunities to use measurement concepts and skills. Teaching about operations with numbers is frequently quite symbolic. Research evidence concerning the learning of number operations indicates that using manipulative materials and models at points in the instructional sequence rather than depending exclusively on symbols enhances learning. It should be noted, however, that the majority of explanatory models and manipulative devices used to demonstrate number concepts and operations are inexorably intertwined with measurement ideas.

The teacher who examines the relative size of the fractions $\frac{2}{3}$ and $\frac{3}{4}$ by having the children construct models of the two fractions in two congruent rectangles is implicitly providing instruction on the comparison idea identified in the previous section. The first-grade child using the number line to find the sum of 17 plus 6 can be having a significant experience with the additivity idea. The use of fraction bars to develop and exhibit the idea of equivalent fractions reinforces the idea of congruence as a fundamental characteristic of area measurement. These are only three examples of how measurement ideas can be used in the teaching of number operations and concepts. The evidence is that judicious use of measurement-based explanatory models for arithmetic concepts is an effective instructional strategy. Understanding of measurement concepts is positively correlated with achievement of computational skills (Babcock 1978, Taloumis 1979).

There is little evidence that using explanatory models based on measurement to teach numerical concepts and operations actually improves performance on measurement tasks. This is because the causal research has not been done. It appears reasonable, however, to assume that the use of measurement concepts to establish numerical concepts should be reinforcing the measurement ideas. For instruction about number using measurement ideas, most curricula leave any objective concerning the learning of measurement concepts implicit. Even without addressing measurement objectives explicitly, I suspect that considerable improvement in the understanding of length, area, and volume can be realized if explanatory models based on these measure systems are used. If mathematics programs would make calculated use of manipulative models to reinforce measurement ideas as well as to establish numerical concepts and skills, performance in both domains should be improved.

More is at stake in using manipulative devices, models, and diagrams to explain operations with rational numbers than simply improving computational proficiency. Instruction should be organized to use models in teaching number concepts in order to reinforce and use measurement ideas as well. This instructional strategy is perhaps the single most effective means of increasing the time given to instruction on measurement, and it has the added advantage of taking no time from that allotted to any other curricular content.

ESTIMATION AND MEASUREMENT

Estimation fits almost everyone's idea of a type of measurement skill that should be developed in school. The fundamental basic character of estimation in the down-to-earth sense of constant application throughout adult life for both personal and vocational use is thoroughly elaborated in the position paper on basic skills produced by the National Council of Supervisors of Mathematics (1978). The data collected in the curriculum preference surveys of the Study of Priorities in School Mathematics conducted by the National Council of Teachers of Mathematics reveal a commitment by nine different populations to the importance of estimation skills. Data pooled from six different populations ranging from elementary and secondary schoolteachers through university mathematicians indicate that close to 90 percent support estimation as one of the important goals of teaching measurement at either the secondary or elementary school level. When queried about the content of instruction, in contrast to the goals, the rate of positive support for estimation was even higher.

The evidence suggests, however, that little teaching of estimation skills actually takes place even though the importance of the topic is so widely recognized. The findings in the *Case Studies in Science Education* (Stake and Easley 1978) document that most teachers of mathematics at both the secondary and elementary levels seldom if ever deviate from the content of the textbook. In an analysis of fourteen recently published and widely used textbook series, I found few lessons that are devoted to estimation skills and techniques at the elementary school level. The "average" series has one lesson on estimation for every sixty-four on other topics (and the computer category system used for the analysis includes approximate computation with measurement estimation). The typical general mathematics texts of the junior and senior high schools devote a smaller portion of the lessons to estimation than is typical of fifth- and sixth-grade texts. And even less attention is given to estimation in the texts of the college preparatory curriculum. Although estimation is almost universally considered important, the evidence is that it is seldom taught.

It could be well argued that estimation is difficult to capture on the written page. The difficulty in "booking" estimation stems from several different factors. First, confinement to the written page restricts activities to one- and two-dimensional objects that are smaller than the page; the learner is thereby restricted to a skill of limited applicability. Second, evidence indicates that estimation skills require constant and frequent practice or they evaporate. Brief, recurrent activities provide practice, but they consume expensive page space. Third, the relative nature of estimation is difficult to capture on the page. Any estimate is correct in an absolute sense; the premium must be on which estimate is better. Textual materials should provide a mechanism for checking an estimate against an accurate measurement in order to give the learner an opportunity to improve the estimate. The process of estimating, checking, and improving the estimate accommodates much more readily to discussion at the moment of estimation than to written description of activities on the printed page. Therefore, the responsibility for providing estimation activities appears more appropriately placed with the teacher than with the text author.

If the responsibilities for designing and implementing estimation activities are the teacher's, what characteristics should these activities have? Unfortunately, little guidance is available for teachers in the professional literature of mathematics education. Regrettably few mathematics educators have written about estimation or conducted research about how estimation skills are best taught.

The following suggestions primarily concern what is to be taught—namely, strategies for estimation. Most textbooks and articles that do consider estimation suggest that it is most effective to estimate and then check the accuracy of the estimate. While good as far as it goes, this practice does not teach any process for making an estimate.

The strategies identified below are the results of an analysis of what people do when estimating. Hildreth (1979) interviewed a sample of junior college students in order to identify the types of estimation strategies they used and to ascertain how accurate they were in making estimates of length and area. The students who exhibited or reported uses of strategies were significantly more accurate estimators than those who did not. This study suggests that instruction should be directed toward teaching strategies.

Following are some strategies used by good estimators. Instruction focused on these strategies is a first approximation for needed curricular emphasis on estimation. Each estimation strategy is labeled by an identifying word and a question that indicates the character of the strategy.

Simple Comparison—Do you know an object that is about the same size? This is the most primitive estimation strategy, thinking of an object that is about the same size, and it is a simple-minded but effective process. For example, an informed guess is made by a girl who observes that a blackboard is about as long as her father is tall and that he is about 6 feet tall. An implication of this strategy is that children need to acquire a set of reference objects for which they know the measures. Use of comparison strategy depends on having prior knowledge about the measures of a set of reference objects. Note that the use of this strategy depends upon having an understanding of the comparison idea for measurement that was described earlier.

Bracketing—Do you know an object that is just a little bigger and one that is just a little smaller than the object being estimated? This estimation strategy is an attempt to "squeeze" the estimated measure between the measures of two known objects. A middle-school student faced with estimating the height of Jon, a new student in the school, might observe that Jon is shorter than his friend Joe who is 5 feet 10 inches but taller than his friend Sue who is 5 feet 6 inches. He can now bracket his estimate between the two known lengths. Bracketing is an extension and refinement of the simple comparison strategy.

Chunking—Do you see a natural way of dividing the object to be estimated in parts for which you know the measure? Partitioning an object into chunks that correspond to the lengths of familiar, known objects makes the estimation task easier. To find the width of the room exhibited in Figure 4-2, a student might observe that the couch "chunk" is about 6 feet long, that the door chunk is about two thirds as big as the couch chunk or 4 feet, and the remaining chunk is also about 4 feet. Adding the estimates of the chunks together gives an estimate of 14 feet for the room width. The heart of the strategy is identifying subportions for which you know the measurement with a fair degree of certainty or for which you can readily employ comparison strategies in order to reduce the size of the portion about which you are uncertain. In general, every chunk you can make a good estimate on reduces the inaccuracy of the overall estimate. This strategy has the added advantage of using and reinforcing the additivity idea—a fundamental property of most systems of measure.

Unitizing—Do you see a way of reproducing a chunk and then counting the reproduction? This estimation strategy is an organized application and refinement of the chunking strategy in which every chunk has the

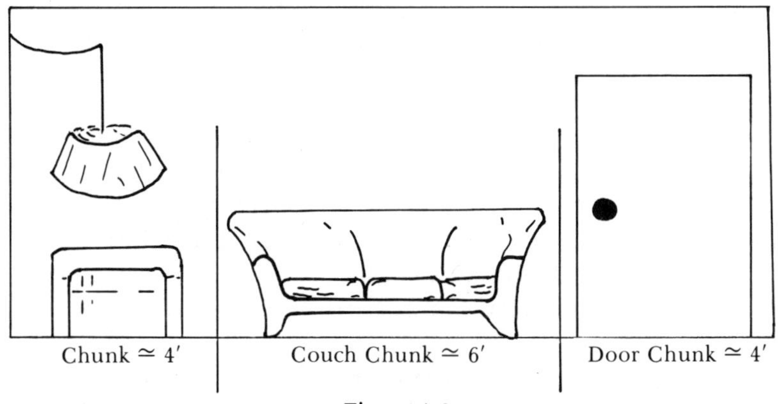

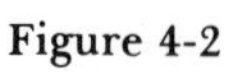

Figure 4-2

An example of how a student might partition a room into chunks for estimating

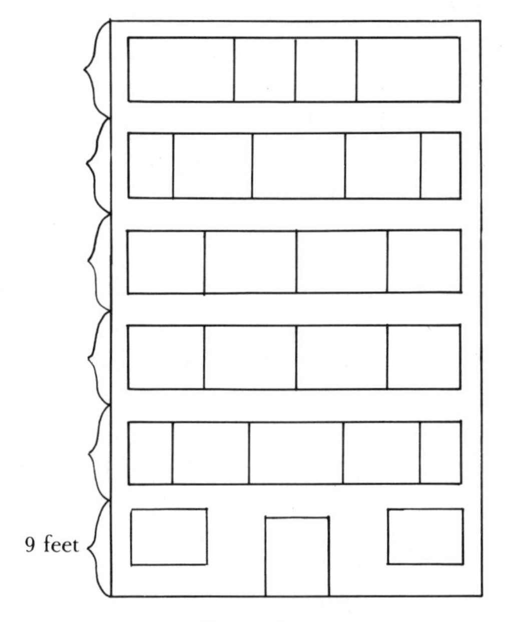

Figure 4-3

An example of unitizing

same measure. Figure 4-3 shows a six-floor building. A student might estimate the height of the building by noting that there are six floors and that each could be used as a unit chunk. Knowing that each floor is about 8 feet from floor to ceiling and that there is about a foot between floors (note the use of simple comparison and of chunking), the student can make an estimate of the height by counting six 9-foot units or by multiplying 6 times 9 feet to get 54 feet.

The unitizing strategy can be employed to estimate the length of smaller objects too. If Bernie can place four toy cars end-to-end in a Christmas mailing carton that is 20 inches long, each car serves as a unit that is about 5 inches in length.

The unitizing strategy reinforces one of the most fundamental characteristics of a measurement system, the concept of unit measure. Since the unit is identified by estimators themselves, the arbitrary nature of a unit of measure is reasserted each time this strategy is employed in estimation. Stressing the unitizing strategy can significantly contribute to the children's understanding of the role of the unit in measurement.

Error Reduction—Can you identify a systematic error in an estimation strategy and compensate for it? A Boy Scout field manual suggests this process for finding the height of a tree based on a unitizing strategy. Affix a temporary marker to the tree a known distance from the ground. Step back a reasonable distance from the tree holding an object such as a pencil in your hand. Align the bottom of the pencil with the bottom of the tree and note the point on the pencil that aligns with the mark on the tree. The portion of the pencil that covers the known distance is the unit for estimation. Find how many pencil lengths it takes to cover the tree as is shown in Figure 4-4 and multiply the unit length times the number of units counted in order to find a reasonable estimate of the tree's height.

Children in the middle school grades can use this estimation strategy, and they can identify a systematic error in the process. Because of the change in viewing angle, the pencil length unit covers a longer part of the tree at the top than at the bottom of the tree. The error gets progressively worse near the top of the tree and becomes more of a problem the closer the estimator stands to the tree. Children can build a compensation into this estimation strategy.

Many estimation processes have systematic errors built in that are similar to this one. Error reduction serves to identify the source of error and compensate for it. It may be inappropriate to call error re-

Figure 4-4

Estimation of the height of a tree

duction a strategy; it is more of an attitude or an awareness factor that the teacher needs to help learners acquire.

These strategies can be extended and elaborated as learners become more sophisticated in estimating. The most powerful extensions employ concepts of proportional thinking. Preliminary evidence indicates that focusing on these strategies provides the estimator with a process that increases the accuracy of estimation. Most text materials do not have the thinking strategies used in estimation as the objects of instruction, but a sound instructional program for children will provide practice as well as help the children refine and incorporate thinking strategies into their processes of estimation. Bright (1976) notes that among the advantages that accrue to giving children experience with estimation is a growth in their ability to understand and use measurement concepts. Given the fundamental relationship between major concepts of measurement and the estimation strategies described above, this is a particularly appropriate observation.

But the identification of estimation strategies offers little by way of suggestions for methods of teaching these strategies. Some characteristics of effectively teaching estimation strategies are described below.

1. Initial instruction should concentrate on establishing or fixing the strategies. Learners need to identify the strategies and discuss them. Discussion appears to be a good way to help the child achieve psychological closure on the strategies. Children need to consider the strategies reflectively, and discussion provides a means of doing this.
2. Activities need to be designed that put a premium on making the error small. Every estimate is correct, but the better estimator is the one who values reducing the error. Activities should allow for the comparison of estimation strategies and give the learner a chance to compare the estimates with accurate measurements made after the estimate was done. I have frequently used a game activity that is scored by having the student find the difference between the estimate and the accurate measurement; the winner of the game activity is the player with the smallest sum of error differences.
3. Estimation activities should be used with various systems of measurement. The strategies were exemplified in terms of length situations. However, the estimation strategies apply quite well to area, volume, weight, and other systems of measure.
4. Regular, frequent practice with estimation activities of short duration scattered over several days of instruction is more effective than using the same amount of time to mass practice in one or two days' worth of activity on estimation. Estimation skills deteriorate if not used, but it does not take much time to keep students at the awareness level needed to maintain the skill.
5. Estimation strategies that feature unitizing are particularly powerful both in terms of providing an accurate estimation process and in reinforcing a basic concept of measurement. However, children have some difficulties in identifying appropriate units. They frequently identify units that are too small and encounter difficulties that are perceptual in keeping track of the counting. Discussion should focus on identifying perceptually natural subdivisions that the child can keep track of in unitizing. Objects will often have features that suggest a natural division into units.
6. Children should see estimation used in a variety of settings in order to appreciate its usefulness. The teacher who finds activities in science, social science, and other settings that will lend themselves to problem solving based on estimation rather than re-

serving estimation activity to mathematics will help the learner appreciate estimation as a general tool for problem solving.

CONCLUDING COMMENTS

Three domains of curriculum change that would improve the effectiveness and efficiency of teaching measurement have been identified:

1. Stressing the mathematical and scientific nature of measurement.
2. Using measurement concepts in teaching about number.
3. Providing extensive experience in estimation.

With the exception of the latter, none of these changes would reduce the time allotted for instruction on other important mathematical topics. The topic of estimation is, however, one of the most important basic skills and is currently not well represented in the instructional programs of most schools. This chapter addresses the question of how much measurement should be taught in the school mathematics program. Clearly schools need to do a better job of teaching measurement, but the competition for time between topics in mathematics will not accommodate a dramatic increase in instructional time on measurement per se. The curricular and instructional tasks described in simplest terms are equivalent to using present instructional time allotments more effectively and efficiently.

REFERENCES

Babcock, Gail R. "The Relationship between Basal Measurement Ability and Rational Number Learning at Three Grade Levels." Doct. diss., University of Alberta, 1978.

Bright, George W. "Estimation as Part of Learning to Measure." In *Measurement in School Mathematics*, edited by Doyal Nelson. Thirty-eighth Yearbook of the National Council of Teachers of Mathematics. Reston, Va.: NCTM, 1976.

Higgins, Jon L., ed. *A Metric Handbook for Teachers.* Reston, Va.: NCTM, 1974.

Hildreth, David J. "The Effects of Estimation Strategy Use on Length and Area Estimation Ability and the Effects of Teaching Estimation Strategies to Fifth and Seventh Grade Students." Dissertation proposal, Ohio State University, 1979.

Hirstein, James J., Lamb, Charles E., and Osborne, Alan. "Student Misconceptions about Area Measure." *Arithmetic Teacher* 25 (March 1978): 10–15.

Lesh, Richard A., ed. *Number and Measurement: Papers from a Research Workshop.* Columbus, Ohio: ERIC Information Analysis Center for Science, Mathematics, and Environmental Education, 1976. ERIC: ED 120 027.

National Assessment of Educational Progress. *Changes in Mathematical Achievement,* 1973-78, Report No. 09-MA-01. Washington, D.C.: U.S. Government Printing Office, 1979 (a).

National Assessment of Educational Progress. *Mathematical Knowledge and Skills,* Report No. 09-MA-02. Washington, D.C.: U.S. Government Printing Office, 1979 (b).

National Assessment of Educational Progress. *Mathematical Applications,* Report No. 09-MA-03. Washington, D.C. U.S. Government Printing Office, 1979 (c).

National Council of Supervisors of Mathematics. "Position Statement on Basic Skills." *Mathematics Teacher* 71 (February 1978): 147–52.

Nelson, Doyal, ed. *Measurement in School Mathematics.* Thirty-eighth Yearbook of the National Council of Teachers of Mathematics. Reston, Va.: NCTM, 1976.

PRISM Project Staff. "PRISM Highlights: An Interpretive Summary from the NCTM Project Priorities in School Mathematics." Mimeographed and circulated at the Annual Meeting of the NCTM, Spring 1980, Seattle.

Stake, Robert E., and Easley, Jack A., Jr. *Case Studies in Science Education.* Champaign-Urbana, Ill.: Center for Instructional Research and Curriculum Evaluation, University of Illinois, 1978. ERIC: ED 156 498–ED 156 512.

Taloumis, Thalia. "Scores on Piagetian Area Tasks as Predictors of Achievement in Mathematics over a Four-Year Period." *Journal for Research in Mathematics Education* 10 (March 1979): 120–34.

5. Knowing Rational Numbers: Ideas and Symbols

Thomas E. Kieren

There has been a long-standing frustration with both teaching and learning about rational numbers. Rational numbers, for the purpose of discussion in this chapter, are those numbers generally called common fractions and decimals. With the current emphasis on metric measure and the widespread availability of calculators, there is a call for the earlier as well as greater use of decimals as opposed to common fractions in mathematics curriculum. It is thought by some that such a change in symbolic treatment (from common-fraction notation to decimal notation) would alleviate the teaching and learning problem and even allow for a reduction of time spent on learning ideas about fractional numbers.

To assess the effect of this suggested change and to discuss curriculum changes with respect to rational numbers, one must ask the following question: "What ideas should a person have about rational numbers?" This is, of course, a philosophical question, but it contains two important implications for teachers and ultimately for learners. The first is that a person's ideas of rational numbers should be *about* something. The first paragraph speculates on the symbolic representation of rational numbers and symbolic knowledge of rational numbers. The question is what is this symbolic knowledge about? The philosopher Bateson (1979) suggests that unless a person's symbolic knowledge is *about* something, the person will inevitably be confused. Is this the state of a student who says ".8 is the same as $\frac{1}{8}$" or "$\frac{9}{3}$ is not equal to $\frac{12}{4}$ because 9 x 3 is not equal to 12 x 4" but does not seem to think that the "nine-thirds" or "twelve-fourths" have other than symbolic meaning?

If one asked mathematicians what rational numbers are *about*, they would say rational numbers are two sided. In one sense they are quantities that can be added and show the structure associated with addition. In another sense they are functions (or relationships between quantities) that can be composed: "If I take $\frac{1}{2}$ of $\frac{1}{4}$ of an amount, I end up with $\frac{1}{8}$ of the original amount." In this sense rational numbers are multiplicative and show the structure associated with multiplication.

The second curriculum impact of the philosophical question derives from the word *should*. The philosopher Margenau (1961) sees knowledge, particularly mathematical knowledge, as constructs or ideas built up by each person. If this knowledge is to be powerful or to free a person to better understand and live in the world, such ideas should be extensive. That is, one's ideas about rational numbers should allow one to work with a wide variety of real-world situations. These ideas should also be *connected*. Ideas about rational numbers should relate to other mathematical ideas—whole numbers, real numbers, transformation geometry, algebra, measurement, and so forth. These ideas should not stand in isolation, for then they are unimportant, unused, and forgotten.

Students often ask "Why should I learn rational numbers?"—in other words, "What are rational numbers *about* in my everyday life?" A fairly typical nonanswer to this question is "Because you'll need them in algebra." But because of their two-sided nature (quantity and relationship among quantity or additive and multiplicative), ideas about rational numbers can provide a person with models of four important real-life situations. While whole-number ideas allow numbers to be applied to discrete phenomena, rational numbers allow for the application of number to continuous quantities. Rational numbers are used in *measuring*—$2\frac{1}{4}$ yards or 3.2 liters. Rational numbers provide answers to division problems—9 buns among 4 people, each gets $2\frac{1}{4}$. (One does not need to say $9 \div 4$ cannot be done or $9 \div 4$ yields 2 with a remainder of 1.) Rational numbers also allow for the application of numbers to the relationship between or comparison of two quantities—the interest rate for a year is the quotient of the cost of the loan divided by the principal and that is the same as $\frac{K}{100}$ or K percent.

Finally, common or decimal fractions are useful in describing part-whole situations.

EXPERIENCE BASE FOR RATIONAL NUMBERS

To be connected and extensive enough to provide models for these real-life situations, a person's knowledge of rational numbers should have three levels: experiential, symbolic, and axiomatic. The first level, the experiential, includes ideas about four categories of experience, or four basic subconstructs of a person's knowledge of rational numbers. These four subconstructs are described here in detail.

Basic ideas about rational numbers develop from several primitive sources. A very early fractional idea is that of half. This idea is about the experience of sharing equally. Certainly the words *one-half* or *half of* are the earliest fraction words a person employs and, in fact, are the main fraction words used even in adult vocabulary. The sharing experience noted above is a precursor to the more general and number-related notion of partitioning—the dividing of a quantity into subparts of equal size or number. It should be noted that partitioning is a learned mechanism and one that a person uses to build up ideas about rational numbers. Based on the primitive ideas of fractional numbers and especially on partitioning, the four experience-based subconstructs can be built.

One subconstruct is that a rational number is a *measure* number. Suppose one wanted to measure the surface area of a region. One would select a convenient unit, cover the region with replicas of the unit, and count such replicas. If this covering is incomplete and a whole unit is larger than the remaining region, the unit is partitioned and replicas of one part are now used to cover the remaining region. If the covering is still incomplete, this part can be further partitioned and the process repeated until the covering is as complete as desired. There is, of course, a correspondence between such physical partitioning and its numerical counterpart (division) as well as its symbolic counterpart. One could say that the area of a region was 1 square meter plus $\frac{1}{2}$ square meter plus $\frac{1}{6}$ square meter or that the area of another region was 1 square meter plus .6 square meter plus .02 square meter plus .005 square meter or 1.625 square meters. The first partitioning noted here is a selective one (first by two, then by three to get sixths). The second partitioning is standardized (by ten, by ten once more to get hundredths, and again to get thousandths). Thus, rational numbers are measure numbers or numbers about measuring experiences. Though it is beyond the scope of this chapter to describe how all aspects of rational numbers can be seen in this light, one should observe that addition of rational numbers has a natural "put together" meaning here. If a triangle has sides measuring $\frac{1}{3}$, $\frac{1}{2}$, and $\frac{1}{4}$ of a unit, one can find the sum by directly measuring the perimeter without considerations of a common denominator algorithm.

As a second subconstruct, rational numbers are *quotients*. Formally this means that rational numbers are solutions to equations of the form, $ax = b$ where a and b are integers and $a \neq 0$. In one sense, as quotient numbers, rational numbers are formal; from this base one can algebraically prove and derive all the important properties of algorithms for rational numbers. In another sense, rational numbers as quotient numbers are very much rooted in experience. The problem of sharing three bars of candy among five persons gives rise to a quotient number, $\frac{3}{5}$. The actual problem-solving process involves partitioning. Rational numbers as quotient numbers allow one to apply number to such partitioning problems.

Rational numbers are also *ratio* numbers. This idea manifests itself, for example, when one looks at a one-to-three mixture of flour and water and observes that one-fourth is flour. This example highlights the fact that if ratio numbers are to have all rational-number properties, a unit must be made explicit. It is interesting to note that within this subconstruct equivalence takes on multiplicative meaning (double the recipe), which has the character of "like" but not "same." The *measures* $\frac{3}{4}$ and $\frac{6}{8}$ of a unit are the same, but the *ratios* 3 to 4 and 6 to 8 are alike but are phenomenally different.

As a final basic subconstruct, rational numbers are *operator* or *mapping* numbers. Such a number can be illustrated by a "$\frac{1}{8}$ operator," which is a mathematical model for a machine that packs gum eight sticks to a pack. Here the number of sticks is paired with a number of packs one-eighth its size. Thus, twenty-four sticks would be paired with three. A projecting machine that projects an image 5.7 times the size of the projected object is again modeled by an operator number. The operator number illustrates two rational-number properties very well. First, it is natural to multiplicatively compose operators. Using the gum example, a "$\frac{1}{8}$ operator" packs four hundred sticks into fifty packs, then a "$\frac{1}{10}$ operator" packs 10 packs per carton; one could say that "$\frac{1}{80} = \frac{1}{8} \times \frac{1}{10}$ operator" packs sticks into cartons. The operator number gives rise to the multiplicative structure of rational numbers.

Within the operator number subconstruct no special status is given to fractional numbers less than one; a "$\frac{2}{3}$ operator" and a "$\frac{3}{2}$ operator" function in the same way. In these ways the operator number is the most algebraic of the basic ideas.

All of the mathematical characteristics of rational numbers can be observed in each of four basic ideas of rational numbers, and each can be thought of as a mathematical variate of the concept. Yet as personal knowledge, each is different. This can be illustrated by an example from the research of Noelting (1978, 1979). He asked a series of questions using either the ratio subconstruct or the quotient number subconstruct while keeping the numerical information the same. A sample of these is given in Table 5-1.

Table 5-1

Sample of questions asked in Noelting's research

Situation 1 (ratio number questions)

Which of the following mixtures has a stronger orange flavor, A or B?

A: One orange concentrate, three water
B: Two orange concentrate, six water

Which of the following mixtures has a stronger orange flavor, M or N?

M: Two orange, three water
N: Four orange, six water

Situation 2 (quotient number questions)

Some cookies are shared among two groups of boys. In which group will a boy get more cookies, A or B?

A: One cookie for three boys
B: Two cookies for six boys

In which group will a boy get more cookies, M or N?

M: Two cookies for three boys
N: Four cookies for six boys

While these questions appear mathematically identical, situation 1 can be thought of as manifesting ratio numbers, while situation 2 exemplifies a quotient number. Noelting found that subjects reacted identically to the two ratio numbers questions but found the first of the quotient number questions easier than the second. In the quotient setting, students seemed to notice unit fractions such as $\frac{1}{3}$ as different from nonunit fractions ($\frac{2}{3}$), while this was not the case in the ratio setting.

Because the four different basic subconstructs can be shown to have similar mathematical properties, one might be inclined to use only one or another in a mathematics curriculum. However, from the learner's point of view, learning about rational numbers is not that simple. As seen in the previous example, situations related to different subconstructs of rational numbers draw qualitatively different responses from students. Further, each basic rational number subconstruct models different applications and allows different extensions to other mathematical ideas. Thus, all four basic experiences should be systematically represented in a curriculum.

There is one further basic idea about rational numbers, the idea of *part-whole* numbers. This subconstruct is nominal in two senses. First, it is the construct from which the name *fraction* arises; this origin is even more evident in a language such as German where the word for fraction is *Bruchzahl* or "broken number." The part-whole construct is nominal in a second more important sense. It is from the part-whole contrast that our language for rational numbers derives. This is true for common fractional notation ($\frac{3}{4}$ or three-fourths) or decimal fractions (.75 or seventy-five hundredths). This language is based on the idea of an *ordered pair*, which comes from the part-whole contrast.

Because the part-whole contrast generates rational number symbolism in a very transparent way, it is not surprising that this idea is heavily emphasized in current curriculums. However, such part-whole experience frequently does not lead to a sufficient understanding of rational numbers. Very early the child is faced with an accomplished partition in which some number of pieces have been shaded or indicated, as in Figure 5-1. The child is then asked to generate part-whole language by counting the number of shaded parts and the total number of parts and reporting this pair of numbers. Such an approach does not relate the act of partitioning to the fractional number. For example, a child under such a curriculum might say "three-fourths" to both situations but not be able to say if the regions are the same size or realize the quantitative significance of "three-fourths."

Further, such emphasis on part-whole without emphasis on the other four basic ideas about rational numbers leads to weak number ideas in other ways as well. In many part-whole treatments, the two situations in Figure 5-2 are treated as almost identical for the learner.

The learner is asked to say "one-eighth" and "three-eighths." The language has the same structure, and the counting to generate it is simple. Still such an approach ignores the fact that one-eighth represents a quantity and that the learner must see this if three-eighths is to have an appropriate quantitative meaning ($\frac{3}{8} = \frac{1}{8} + \frac{1}{8} + \frac{1}{8}$). It also

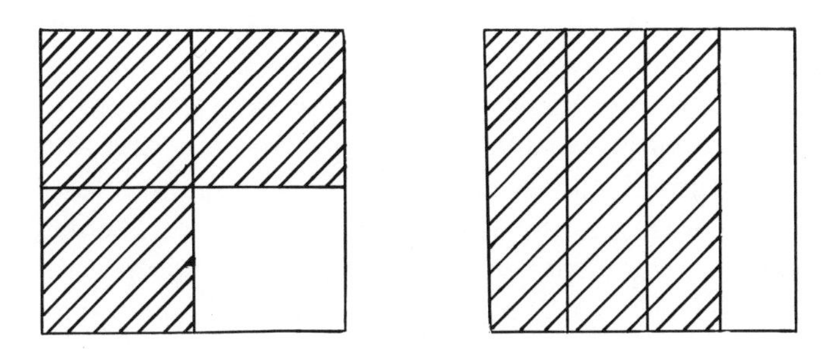

Figure 5-1
Two models of three-fourths

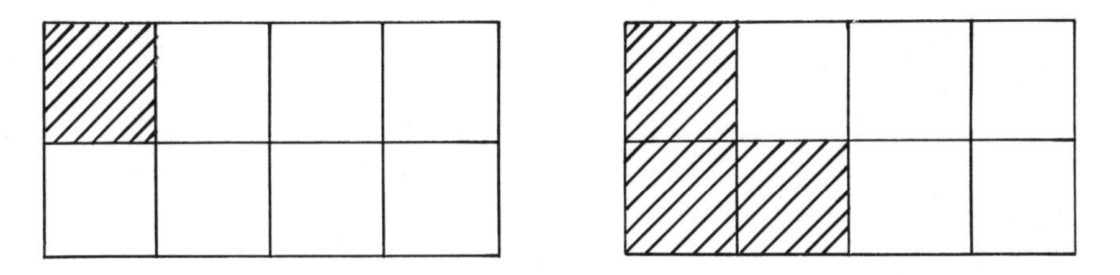

Figure 5-2
Models of one-eighth and three-eighths

ignores the fact that $\frac{1}{8}$ is much simpler for a child than $\frac{3}{8}$ when the subconstruct is a quotient number. Thus, while important for language generation, the part-whole number idea is an inappropriate base for the development of number ideas without connection to the other four subconstructs described.

BUILDING UP IDEAS ABOUT RATIONAL NUMBERS

While it is true that general cognitive development is important for the development of various number ideas in a person, a person's use of a number of learned tools is also very important. For example, various kinds of counting strategies are used as children build up, examine, and use whole number ideas.

With respect to building up ideas of rational numbers, there are a number of thinking tools that would be useful if available. Two of these, and perhaps the most important, are *partitioning* and *equivalence*.

Partitioning, the division of a quantity into a number of equal-sized pieces, has been discussed in many contexts in this chapter. Partitioning develops from informal sharing situations as well as from halving activities. Two early specific forms of partitioning are "dealing out" and repeated binary division (in half, in half again, and so forth).

In its early form, equity among parts is the central concern of the partitioning act. Later partitioning is related to quantity and number (the size of the part is related to the size of the object and the number of parts). Still later, partitioning is related to the formal number activities of division and factoring. Finally, a person is able to compare rational numbers or add them by being able to compare two partitionings or factorings. This can be seen in the work of someone who can "reduce" a fraction, generate a set of equivalent fractions, or exhibit "common-denominator" thinking.

In many curriculums, equivalence is treated very early in a formal sense: $\frac{a}{b}$ is equivalent to $\frac{c}{d}$ if a x d = b x c. Yet, if equivalence is to be a problem-solving, idea-building tool, it, too, must be *about* something for a person. As an idea, equivalence grows from an informal to a formal level. One early form is additive and quantitative in nature. A child, knowing nothing about common denominators, is asked "How much is $\frac{3}{4}$ plus $\frac{1}{2}$?" and replies, "Well, $\frac{1}{2}$ is $\frac{1}{4}$ plus $\frac{1}{4}$, so $\frac{1}{4}$ takes you up to 1 and you have $\frac{1}{4}$ more; so it's $1\frac{1}{4}$." This early form of equivalence is especially useful in developing and using rational numbers as measure numbers.

In general, early equivalence ideas seem to be "works like" ideas. After seeing several instances of a $\frac{3}{4}$ operator, a junior high student is asked what the output of a machine ($\frac{3}{4}$ machine) will be if the input is 24. The student says, "You divide by 4 then multiply by 3 and get 18." This person is seeing the rational number $\frac{3}{4}$ in a rule or process sense. It should be noted that this rule is proportional or multiplicative in nature (Kieren and Southwell 1979).

Later, equivalence focuses on pairs. A student is asked to generate a pair like $\frac{3}{4}$ with a numerator of 24 and says, "I know $\frac{12}{16}$ is like $\frac{3}{4}$, therefore (2x12)/(2x16) or $\frac{24}{32}$ must be a pair." Still later persons recognize that equivalence means an entire class of fractions. Even later, the formal mathematical idea of rational numbers as equivalence classes of pairs of integers may develop.

It should be noted that premature introduction of the cross multiplication rule ($\frac{a}{b}=\frac{c}{d}$ if ad=bc) or the idea of a set of equivalent fractions misuses their powerful tool of equivalence in rational number development.

ON SYMBOLS

Let us now reconsider the call for greater and earlier emphasis on decimals as opposed to common fractions in the mathematics curriculum. Frequently this is associated with a call for research on the relative effects of common fraction–decimal fraction versus decimal fraction–common fraction instructional sequences or research comparing the efficiency of computation with common as opposed to decimal fractions.

The preceding analysis of the basic ideas of rational numbers and of partitioning and equivalence makes such curriculum and research questions inappropriate. One appropriate question to ask is which symbol form is most useful at a particular stage of development of knowledge about rational numbers. A second question is the manner in which symbolic forms can be related to the various basic ideas and building tools of rational numbers.

There is evidence that language is important in understanding rational numbers. Gunderson and Gunderson (1957) and Payne (1976) found that before presenting fractions in any numerical form, it was important to use the word form ("three-fourths" as compared with "$\frac{3}{4}$") for a period of time. In each of the four basic contexts of rational numbers, there is an early informal language that can be used. These are briefly shown in table 5-2.

Table 5-2

Informal language of basic contexts of rational numbers

Context	Language	Symbol
measure	three-fourths of a unit	$\frac{3}{4}$
quotient	three divided by four	$3 \div 4$
ratio	three to four	3 : 4
operator	three for every four	3 for 4

This table suggests that a variety of informal languages and informal symbols be used in that part of the curriculum involving rational or fractional numbers. It further suggests that a person who has knowledge of rational numbers must be able to judge the connotation of the fraction $\frac{3}{4}$. It also points to the problem of connecting decimal notation, in this case .75, to the various basic contexts of fractions.

There are distinct differences between using the language of decimals and that of common fractions in the various contexts. For example, if one asks how much each person gets if two pizzas are divided among three persons, two-thirds ($\frac{2}{3}$) is a natural answer reflecting the partitioning of the pizzas in threes and giving two pieces to each person. The decimal response ($.\overline{6}$ or .6666 . . .) is phenomenally more complex. It can be thought of as representing the procedure of dividing the two pizzas into tenths, giving 6 tenths to each person, dividing the remaining pieces by ten, giving 6 hundredths to each and repeating. In this context "$\frac{2}{3}$" is a simpler and better symbolism than "$.\overline{6}$." In measuring molecular or atomic particles, the relationship between decimal notation, scientific notation, and significant digits makes decimal notation the obvious symbolism.

Different notations seem more appropriate for the different basic rational number ideas. With measure numbers either language form is appropriate. Because measure often involves comparison, the decimal notation allows for easy comparison and at times may be more appropriate. In the operator construct, describing the transformation of 24 input items for 18 output items as a "3 for 4" exchange or a "$\frac{3}{4}$" transformation relates to the context better than ".75." The notation of common fractions is also suggestive of a simple "divide by 4 and multiply by 3" equivalence rule. Thus, in curriculum development the notation of choice or emphasis would differ from construct to construct.

Common fractional and decimal notations also connote different conceptions of partitioning. The former ($9\frac{1}{3}$, $\frac{3}{20}$, $\frac{5}{8}$) suggests that the number of partition depends on the situation. The latter ($.\overline{3}$, .15, .625) connotes repeated partitioning by ten. This latter standardized partitioning is a special case and is likely more sophisticated than partitioning in general. In using partitioning to build ideas about rational numbers, the common fraction language may be simpler or more appropriate than the decimal language.

Because symbolic knowledge should be about something, decimal symbolism should reflect certain ideas and experiences. In developing the decimal system of whole numbers, the idea of grouping by ten, collecting tens of "ones," "tens," and the like, is central. The notation of decimal fractions reflects a different process. The basic experience as noted above is partitioning into ten parts. There must be an abundance of such experience, and because dividing a set or a region into ten parts is a clumsy task, such experiences are not easy to develop in a curriculum. It may be that such partitioning would have to be an idea generalized from other simpler repeated partitionings (by twos, threes, and so forth). However, if in metric measuring the meter is used as the original physical unit in the curriculum, then decimeters as tenths (.1), centimeters as hundredths (.01), and so forth can serve as basic decimal experience. (Most curriculums use centimeters or decimeters as the basic physical unit handled by the child. This leads to a meter being seen as ten decimeters, which is not useful in decimal fraction development.)

Both decimal and common fraction notation have important places in the symbolic knowledge of a person. Decimal notation allows for simple and mathematically important computational algorithms. The generalization from whole number to decimal fraction algorithms makes for obvious efficiency. The decimal notation also allows for the computation of irrational numbers as a sequence of rational approximations; for example, the square root of 2 may be approximated by the sequence 1.4, 1.41, 1.414, and so on.

The idea of equivalence and equivalence class in decimal terms is almost an empty one (for example, .75, .750, .7500, .75000 . . .). Common fraction notation, however, allows for generation of equivalent pairs and the notion of equivalence class. For example, $\frac{3}{4}$ is equivalent to $\frac{9}{12}$ and the equivalence class of $\frac{3}{4}$ is the set of all equivalent fractions ($\frac{3}{4}, \frac{6}{8}, \frac{9}{12}, \frac{12}{16} \ldots$). Common fractions are also basic to the development and application of rational forms used in many everyday formulas such as $r = \frac{d}{t}$ or $M = \frac{e}{c^2}$.

A NOTE ON SEQUENCE

It is beyond the scope of this chapter to discuss a possible best curriculum for rational numbers, yet the analysis presented in this chapter is suggestive of curricular implications. It is not hard to conceptualize simple experiences relating to any of the four basic rational number ideas: measure numbers, quotient numbers, ratio numbers, and operator numbers. The quotient number experiences (for example, sharing four pizzas among seven children) appear most naturally related to partitioning. Measure number experiences both generate and make early use of the notion of equivalence. In addition, they provide an experience base for decimal symbolism and the associated "by tens" partitioning. Thus it appears that measure and quotient number experience might form an early base for rational number curriculums. This, however, is an open development, and four ideas are feasible beginning points. It is more a question of the optimum development of students' personal knowledge of rational numbers.

Finally it should be recognized that both algorithms and structural or axiomatic knowledge of rational numbers must be built upon the basic ideas about rational numbers. Within each of the four basic contexts, ideas of operations and equivalence can be developed informally using an informal language. But the development of and understanding of standard algorithms and more formal equivalence ideas must follow.

CONCLUSION

Knowledge of rational numbers, which is extensive and connected, is a complex multilevel phenomenon. It should be built up by a person from experience with a variety of basic ideas of rational numbers. A person uses mechanisms such as partitioning and equivalence, which themselves become more elaborate and formal, to build up this knowledge of rational numbers.

Knowledge of rational numbers is an important tool for persons in the everyday adult world as well as in other academic pursuits. Because children have only limited out-of-school experiences with rational numbers and these mechanisms, the curriculum must be designed to provide all of the basic number experiences and symbolic experiences useful in building a sound personal number construct. It is important in analyzing the current curriculum and in building new ones that such experiences reflect the informal as well as the formal knowledge of rational numbers.

REFERENCES

Bateson, Gregory. *Mind and Nature: A Necessary Unity.* New York: Dutton, 1979.

Gunderson, Agnes G., and Gunderson, Ethel. "Fraction Concepts Held by Young Children." *Arithmetic Teacher* 4 (October 1957): 168–73.

Kieren, Thomas E. "The Rational Number Construct: The Elements and Mechanisms." In *Recent Research in Number Learning*, edited by Thomas E. Kieren. Columbus, Ohio: ERIC Information Analysis Center for Science, Mathematics, and Environmental Education, forthcoming.

Kieren, Thomas E., and Southwell, Beth. "Rational Numbers as Operators: In Development of This Construct in Children and Adolescents." *Alberta Journal of Educational Research* 25 (December 1979): 234-47.

Margenau, Henry. *Open Vistas: Philosophical Perspectives of Modern Science.* New Haven, Conn.: Yale University Press, 1961.

Noelting, Gerald. "Constructivism as a Model for Cognitive Development and (Eventually) Learning." Québec: *Ecole de Psychologie, Laval Université*, 1978.

Noelting, Gerald. Untitled paper presented at the Annual Meeting of the National Council of Teachers of Mathematics, Boston, 1979.

Payne, Joseph N. "Review of Research on Fractions." In *Number and Measurement: Papers from a Research Workshop*, edited by Richard A. Lesh. Columbus, Ohio: ERIC Information Analysis Center for Science, Mathematics, and Environmental Education, 1976.

6. Probability and Statistics: Today's Ciphering?

Jane Donnelly Gawronski and *Douglas B. McLeod*

Most of us rarely have a need to solve a quadratic equation or to prove that two triangles are congruent, but many of us have spent substantial amounts of time (and rightly so) learning how to perform these tasks in secondary school mathematics courses. These skills are necessary in order to learn and apply geometric and algebraic concepts in more advanced mathematics. At the elementary level, substantial blocks of time are invested in teaching children to solve fairly complicated problems in long division. Yet today most of us would avoid such problems, unless we had a calculator readily available. Since the above are not skills most of us use daily, the question of what mathematical skills we do need and apply with some frequency outside the classroom must be asked.

Mathematical skills—in addition to the obvious computation needed in making change, estimating amounts, and balancing checkbooks—are certainly essential. For example, the average citizen has to deal almost daily with concepts from probability and statistics. As the report of the National Advisory Committee on Mathematical Education points out (NACOME 1975), these concepts are needed to deal with weather reports, opinion polls, advertising claims, national health problems, nuclear accidents, unemployment, sporting events, and a host of other concerns. Furthermore, these concepts occur in the lives of everyone—in daily conversation, on television, in the newspapers. For example, the current energy crisis has prompted many newspapers and magazines to illustrate supplies of crude oil, natural gas, and/or coal with a table or graph. This provides the reader with a visual display of information that is easily interpreted. This is an application of how statistics are used in everyday publications. However, most of us have studied these concepts only briefly, if at all, in our elementary and secondary school mathematics programs.

The purpose of this chapter is to describe the importance of probability and statistics in the mathematics curriculum of the elementary and secondary schools. We will briefly review some of the important curriculum recommendations on this topic and discuss current classroom practices and student performance. Then we will make some recommendations on what the future will or should hold for instruction in probability and statistics.

CURRICULUM RECOMMENDATIONS

The NACOME (1975) report discusses the recommendations made by various groups concerning the inclusion of probability and statistics into the mathematics curriculum. The most important of these groups is the Joint Committee on the Curriculum in Statistics and Probability of the American Statistical Association and the National Council of Teachers of Mathematics. This committe found that the curriculum was particularly weak in its attention to statistical concepts and that there was a lack of suitable texts. In response to their findings, they sponsored two major publications. The first (Tanur et al. 1972) is a volume of essays that provides a broad and nontechnical introduction to applications of statistics and probability. The purpose of this volume is to persuade teachers, parents, and administrators that statistics should be a central part of the mathematics curriculum. The second publication (Mosteller et al. 1973) is a four-volume work designed to present statistical techniques in terms of real-world problems that show the interdisciplinary nature of statistical work.

The NACOME (1975) report applauds these efforts at improving secondary school offerings in statistics and recommends that more be done in order to integrate probability and statistics into the elementary schools as well. One effort (Romberg et al. 1974–76), an elementary school mathematics program developed at the University of

Wisconsin, includes sections on probability and statistics. In this program, basic concepts of probability and statistics are taught through problem-solving strands that deal with topics of interest to children. Students learn how to describe, organize, display, and interpret data as they construct a foundation of concrete experiences upon which they will later build more abstract concepts.

Other curriculum development efforts have attempted to follow the NACOME recommendations for the elementary school too. For example, the Mathematics Resource Project (1978), developed at the University of Oregon, provides supplementary material in probability and statistics for elementary and middle-school teachers and students. Other programs, such as Unified Science and Mathematics for Elementary Schools (USMES 1973), also have a substantial component on probability and statistics.

The NACOME (1975) report also recommends further development of probability and statistics courses at the secondary level. It suggests a statistics course that would be available to all students, including those with no background in algebra, as a valuable course especially for students who will not go on to college. Such a course is still quite rare, but sample offerings do exist. Gallagher (1979), for example, reports on a somewhat more advanced course for non-college-bound students. In her course, the emphasis is on the development of concepts with the study of relevant problems. Students proceed at a rather slow pace but study fairly difficult concepts, including inferential statistics. Positive responses from students are reported.

Further recommendations by the NACOME report continue the earlier suggestion of other groups, that advanced secondary school students have a statistics course available for their senior year. This recommendation, now more than twenty years old, has yet to be widely implemented. The NACOME report also emphasizes the importance of interdisciplinary courses that use statistics and computers to help demonstrate the typical ways in which social science problems may be addressed.

The Huntington II Project (1973), supported by the National Science Foundation, developed several computer-based simulations in the areas of social studies, biology, and physics. In the simulations students are asked to respond to a series of questions concerning the particular area under investigation. For example, in the water pollution area, students are asked to select the kind of body of water, the rate of temperature at which sewage will be added, and the kind of filtration system. The computer then displays the change in oxygen content of this body of water for several days in both table form and graph form. Students can thus visually and dramatically see the results of the decisions they made. They can also rerun the same simulation many times to research the differences when one of the parameters, say temperature, is changed.

CLASSROOM PRACTICE AND STUDENT PERFORMANCE

The most important curriculum recommendations of the past twenty years have all referred to the need for probability and statistics to be included in the elementary and secondary school mathematics programs. Most programs do include at least some topics from this area, even though there is a tendency to present only short sections on probability and statistics that can be omitted easily. Frequently, for example, the last chapter of a ninth-grade algebra book will deal with probability and statistics.

Some data on classroom practices have been collected in recent years, and generally they indicate that probability and statistics are frequently ignored in the classroom. If the topic is treated at all, it is covered only briefly and is frequently used as a filler, such as during the last week of school before a vacation. In a survey conducted for the NACOME (1975) report, it was found that 64 percent of elementary schoolteachers provided little or no instruction in probability and 52 percent did little or nothing on graphs and statistics. At the secondary level, separate courses on probability and statistics are offered at only a small proportion of schools. In a recent survey in Pennsylvania, Callihan and Bell (1977) found that about 15 percent of the schools offered probability and statistics as a separate course, mainly at an advanced level as an elective for seniors. Data from the NACOME report indicated an even smaller percentage of schools nationwide would have such a course. And the notion of having a separate course for non-college-bound students has not yet been adopted in even a minority of these schools. Courses like that described by Gallagher are still quite rare.

Research studies suggest that elementary school students have little difficulty learning to solve simple problems in probability and statistics. But the results of the National Assessment of Educational Progress (NAEP) indicate wide variation in students' performance on different types of problems. About 75 to 90 percent of 17-year-olds and adults could solve a simple problem that asked for the arithmetic average (or mean) of a set of numbers, and almost half of the 13-year-olds could solve such a problem. Finding other averages, however, including the median or a weighted mean, was much more difficult.

NAEP also included a number of probability problems. Performance on these problems was uniformly poor. For example, only 11 percent of the 13-year-olds could apply the definition of probability to a simple urn problem. And most 17-year-olds and adults were unable to solve a simple problem on the probability of getting heads on the toss of a coin (after three earlier tosses of the same coin). These poor performances indicate that concepts from probability and statistics are largely ignored in most mathematics courses in both elementary and secondary schools. These results were from the first national mathematics assessment, and there was little change by the time of the second assessment, as noted in Chapter 13 of this volume. Clearly, much more time and thought needs to go into the development and implementation of concepts of probability and statistics at all grade levels.

RECOMMENDATIONS

In his influential book, *The Process of Education*, Bruner (1960) outlined some important points for curriculum developers. He emphasized the need for presenting fundamental ideas of the discipline to students at an early age and then returning to these ideas repeatedly in a spiral curriculum. As the students mature, these fundamental ideas should be presented in more and more complete forms, depending on the capabilities and the developmental level of the students. In our recommendations on probability and statistics, we would like to follow the writing of Heitele (1975) and his suggestions for fundamental ideas (in the sense of Bruner).

What are the fundamental ideas of probability and statistics that should form the structure of the curriculum? It appears to us that there are about six fundamental ideas most significant for students in the elementary and secondary schools. First, students need to be introduced to a variety of ways of describing and representing data. They need to construct and interpret graphs and tables that represent raw data. They need to represent data with various kinds of averages or measures of central tendency, including mean, media, and mode.

They need to be able to identify the importance of scatter, or the variance, as another way to distinguish one set of data from another.

In the early grades these experiences may be the counting and display of birthday months, favorite ice cream flavors, shoe colors, or even time spent watching TV each week. These activities lead naturally to discussions of "most," "least," range, and eventually the terminology and sophistication of measures of central tendency. This provides children the opportunity to organize, display, and interpret data they collected, and this practical approach is a use of statistics although not usually (and properly so) introduced as a lesson on statistics.

Second, students need to have a variety of experiences with the notion of a sample. Frequently we are most interested in characteristics of an entire population, but the only way to gather information on that population is by sampling and then estimating population characteristics from sample characteristics. Classroom experiences may include estimating the number of yellow jelly beans in a jar of jelly beans by collecting a sample and counting the yellow ones or estimating the blades of grass in a square meter by counting the blades of grass in a square centimeter. These activities are really not very different, conceptually at least, from those of a research microbiologist. And they do provide the necessary concrete experiences with fundamental notions of probability and statistics.

A third and related fundamental idea is the notion of randomness, or of random variables. The importance of having a random sample, rather than just a sample, is frequently not appreciated by students. Experiments similar to the ones previously identified can be repeated with some changes. Suppose only jelly beans from the top layer are selected for the sample—is this random? If the jelly beans are stirred and a sample taken with a dipper, are the results the same? Which random sample is a random distribution? Sampling procedures used to predict elections or to determine TV ratings or consumer response to a new product should be identified and discussed here.

In probability, the notion of a sample space, or the collection of all possible outcomes, forms our fourth fundamental idea. Once a sample space is identified, the calculation of simple probabilities is frequently quite easy. For example, the sample space for tossing one coin would include a head and a tail, but the sample space of tossing two coins at the same time would include head-head, head-tail, tail-head, and tail-tail. Since some events are more likely to occur than others, it is natural to order events on their likelihood of occurrence and to assign a number to an event.

The process of assigning a number (from one to zero) to an event, where the number is the probability, is a fifth fundamental idea. When we tossed one coin, there were only two possible outcomes, and each one occurred once. Thus the probability of tossing a head is $\frac{1}{2}$ and the probability of tossing a tail is $\frac{1}{2}$. When we tossed two coins at the same time, there were four possible outcomes. Since the combination head-head occurred only once, its probability is $\frac{1}{4}$ and likewise for the tail-tail combination. However, a combination in which there is one head and one tail occurs twice so that probability is $\frac{2}{4}$ or $\frac{1}{2}$.

Finally, one of the most difficult of the fundamental ideas is the notion of independence. Students frequently have a view of the world that assumes similar events must be related, or dependent. If three tosses of a coin result in two heads, as proposed in one NAEP problem, many students estimated the probability of heads on the next toss as $\frac{2}{3}$ rather than $\frac{1}{2}$. Helping students obtain a clear understanding of the independence of events requires very careful curriculum development.

A detailed discussion of these and other fundamental ideas of probability and statistics appears in Heitele's paper. Although an emphasis on these fundamental ideas is only one aspect of curriculum development, we believe it is important. Much of the current curriculum in probability and statistics can be characterized as informal and unstructured experiences at the elementary level, followed by dull, dry textbook exercises at the secondary school level. In neither instance does one have the feeling of a unified, well-developed set of curricular ideas on which successful instruction could be based.

Instruction in probability and statistics is currently undergoing major changes at the university level. It is assumed that all students will have access to a calculator, and most courses also make use of computer programs that perform all kinds of data analyses. As the NACOME report suggests, the availability of calculators and computers will continue to increase, and future curriculum development needs to take account of that fact. In addition to computers that come equipped to run all kinds of data analyses, we can expect more and more use of various kinds of computer-generated graphical displays. These new graphical techniques should prove particularly helpful in instruction as well as in interpreting the data.

Finally, we believe that the main obstacle to more adequate treatment of probability and statistics in the elementary schools is our lack of communication about the fundamental ideas to the general public. There is an underlying consensus among many groups that probability and statistics are somehow frills and not basic skills. We hope that the recent statements of the National Council of Supervisors of Mathematics (1978) about the importance of developing basic skills in probability and statistics will be widely heard and heeded.

REFERENCES

Bruner, Jerome S. *The Process of Education.* Cambridge, Mass.: Harvard University Press, 1960.

Callihan, Hubert D., and Bell, Frederick H. "Probability and Statistics in High Schools?" *School Science and Mathematics* 77 (May–June 1977): 418–26.

Carpenter, Thomas P.; Coburn, Terrence G.; Reys, Robert E.; and Wilson, James W. *Results from the First Mathematics Assessment of the National Assessment of Educational Progress.* Reston, Va.: NCTM, 1978.

Gallagher, Joan. "Statistics for the Non-College-Bound Student." *Mathematics Teacher* 72 (February 1979): 137–40.

Heitele, Dietger. "An Epistemological View on Fundamental Stochastic Ideas." *Educational Studies in Mathematics* 6 (July 1975): 187–205.

Huntington II Project. Maynard, Mass.: Digital Equipment Corporation, 1973.

Mathematics Resource Project. *Statistics and Information Organization.* Palo Alto, Calif.: Creative Publications, 1978.

Mosteller, Frederick; Tanur, Judith M.; Kruskal, William H.; Link, Richard F.; Pieters, Richard S.; and Rising, Gerald R., eds. *Statistics by Example: Exploring Data, Weighing Chances, Detecting Patterns, and Finding Models.* 4 vols. Reading, Mass.: Addison-Wesley, 1973.

National Advisory Committee on Mathematical Education (NACOME). *Overview and Analysis of School Mathematics.* Washington, D.C.: Conference Board of Mathematical Sciences, 1975.

National Council of Supervisors of Mathematics. "Position Statement on Basic Skills." *Mathematics Teacher* 71 (February 1978): 147–52.

Price, Jack; Kelley, John L.; and Kelley, Jonathan. "'New Math' Implementation: A Look inside the Classroom." *Journal for Research in Mathematics Education* 8 (November 1977): 323–31.

Romberg, Thomas A.; Harvey, John G.; Moser, James M.; and Montgomery, Mary E. *Developing Mathematical Processes.* Chicago: Rand McNally, 1974, 1975, 1976.

Tanur, Judith M.; Mosteller, Frederick; Kruskal, William H.; Link, Richard F.; Pieters, Richard S.; and Rising, Gerald R., eds. *Statistics: A Guide to the Unknown.* San Francisco: Holden Day, 1972.

Unified Science and Mathematics for Elementary Schools (USMES). Newton, Mass.: Education Development Center, 1973.

7. Geometry: What Shape for a Comprehensive, Balanced Curriculum?

Phares G. O'Daffer

In the fourth century B.C., the Greek philosopher and teacher Plato (492–348 B.C.) carved the inscription "Let no one ignorant of geometry enter my doors" above the entrance to his academy. Plato's influence on thought in a variety of disciplines has been continuous for more than 2,400 years, and yet we still ask these basic questions: What is geometry? Why teach geometry? and, if we teach geometry, What should be taught? and How should it be taught?

The answer to the first of these questions comes from the realm of philosophy and mathematics. The answers to the other questions hinge on current theories of teaching and learning and our perception of the nature of an appropriate curriculum for the schools. In this chapter some of the current thinking on these questions will be reviewed in order to stimulate further thinking on the role of geometry in a balanced curriculum.

WHAT IS GEOMETRY?

As we consider the nature of geometry, it is interesting to imagine when, where, why, and how humans first became involved with geometric ideas. Almost involuntarily our thoughts turn to objects such as seashells, sunflowers, pine cones, and honeycombs, snowflakes, stars, and spiderwebs. In this context, it becomes clear that geometric form and structure have always permeated the universe and that humans have been immersed in a geometric environment from the very beginning. As early inhabitants observed the world around them, they began to abstract geometric ideas and draw pictures to represent them. Later it became useful to name them, to define them more accurately to enhance communication, and to study the more complex relationships between these abstracted ideas. Finally, these refined ideas were reapplied to the real world in simple as well as sophisticated situations.

To fully understand the nature of geometry, one must realize that it has many dimensions. It can be studied by focusing on its origins in nature and the imitations in human-made objects, or it can be studied as a logical, organized body of knowledge, much as presented by Euclid in 300 B.C. in his *Elements of Geometry*. At the highest levels, it can be studied as a formal, axiomatic structure. This level is indicated in Einstein's reference to non-Euclidean geometry in a lecture given in 1921, "Geometry and Experience," when he said, "To this interpretation of geometry I attach great importance, for should I have not been acquainted with it, I would never have been able to develop the theory of relativity."

Geometry can also be experienced in an aesthetic sense. It appears that humans differ from other animals in that they are able to create and admire beautiful objects. Geometers join the ranks of artistic humans in that they are able to create and admire beautiful ideas. Often these ideas are expressed in the actual creation of paintings, sculptures, and other art objects. Much can be learned about geometry by studying the creation of artists. Also, the recreation potential of geometry has been known and exploited for centuries. When we complete a geometric puzzle perhaps we add, ever so slightly, to our understanding of geometry.

While it is impossible to give an in-depth analysis of the question "What is Geometry?" in this chapter, we can oversimplify and state that geometry is the study of space and spatial relationships.

To be complete, this definition would require significant mathematical, philosophical, and pedagogical extension and interpretation. In this context perhaps it will suffice to say simply that geometry emanates from the physical world, involves patterns, can be beautiful, and has applications in the real world. It can be logically challenging, is a fertile vehicle for creativity, and it can be fun.

PRELIMINARY OBSERVATIONS ON TEACHING AND LEARNING GEOMETRY

P. M. van Hiele, as a result of a study on students' mental development in geometry, identified the following five levels (abbreviated from Wirszup 1976).

Level one: The student learns some vocabulary and recognizes a shape as a whole (for example, the student can differentiate between various geometric figures, such as squares, parallelograms, rectangles, triangles, and can associate names with basic geometric figures).

Level two: The student begins to analyze figures (for example, a triangle has three sides, an isosceles triangle has two sides exactly the same length).

Level three: The student logically orders figures, understands interrelationships between figures and the role of definition (for example, relationships such as parallel, perpendicular, congruent, similar; the idea that a definition gives the necessary and sufficient attributes for a particular classification of a figure).

Level four: The student understands the significance of deduction and the role of postulates, theorems, and proofs. (These are often the major goals of plane geometry in the secondary school.)

Level five: The student attains an understanding of rigor and is able to make abstract deductions. (This understanding may not be attained until a student encounters further work in geometry at an upper secondary or university level.)

This description, although by no means complete, provides a helpful framework for discussing the geometry curriculum.

Geometry experiences in the elementary and junior high school are focused primarily on helping children attain levels one, two, and three. It is not uncommon to find that even though the text material used by an elementary schoolteacher includes a number of geometric topics, the teacher will skip much of it. Many teachers who omit geometry are often operating on premises established when they studied geometry in high school. They seem to feel that geometry is a rigorous, proof-oriented type of subject that would be uninteresting and difficult for them and the children in their classes. This emphasis on the deductive process is sometimes even reinforced in preservice courses for elementary teachers and has often led teachers to a narrow view that has limited their ability either to view geometry creatively or to enjoy geometric activities.

Yet an analysis of geometry activities available for elementary and junior high schoolchildren indicates that they are presented informally, without proof, and often utilize active involvement with concrete materials. Children begin to learn *concepts* by observing natural objects in their environment and classifying them or geometric representations in terms of selected characteristics such as corners, size, number of sides, number of corners, and so forth.

Later, as children progress to van Hiele's level two and level three, they learn geometric construction *skills* through active involvement with tools such as rulers, compasses, plexiglass pieces (Miras), and protractors (O'Daffer and Clemens 1976, p. 386). Often at the junior high school level students can be provided with opportunities to discover important geometric *generalizations*.

Many teachers are surprised to learn that specific emphasis on complicated terminology and formal definitions need not be given until late junior high or early secondary school. Proof almost always comes later, in the secondary school, where students are helped to attain van Hiele's level four and are encouraged to reach toward level five.

Thus, in geometry, as in other areas of the curriculum, we teach concepts, generalizations, facts, and skills. Concepts are often initially developed through extensive experience with physical materials and are ever broadened and expanded as the child proceeds to each new level. The generalizations, later to be called the theorems of geometry, can be discovered initially through inductive reasoning, aided by the use of models, cutouts, pictures, measuring, constructions, and other manipulative aids. Later, at the secondary level, students use logic, postulates, definitions, and so forth, and prove that these generalizations are indeed true. The facts of geometry come from selecting certain attributes of figures or certain generalizations that are deemed important enough to remember. The skills of geometry involve, at the lower level, activities such as performing ruler and compass constructions. Later skills include applying formulas to find the area of a figure. Finally, the students gain facility with higher-level skills such as proving theorems and solving geometric problems.

If teachers and others interested in curriculum development are fully aware of the possibilities for teaching geometry creatively and feel that geometry can be presented to fit the thinking and learning patterns of the student at all levels, the task of establishing a well-planned role for geometry in the curriculum is attainable.

WHY TEACH GEOMETRY?

In the preface to his book *First Lessons in Geometry*, published in 1854, Thomas Hill makes the following remarks:

> I have addressed the child's imagination rather than his reason because I wish to teach him to conceive of forms. The child's powers of sensations are developed before his powers of conception and these before his reasoning powers. I have therefore avoided reasoning and simply given interesting geometric facts, fitted, I hope, to arouse a child to the observation of phenomena and to the perceptions of forms as real entities. (Reeve 1930, p. 10).

From that time onward organized groups such as the National Education Association Committee on Elementary Education in 1890, the Committee of Ten on Secondary School Studies in 1893, the National Education Association Committee of 15 on Geometry, the Cambridge Conferences of 1963 and 1967, and, finally, the National Council of Supervisors of Mathematics in 1978 have recommended that informal geometry be taught in the elementary school starting with kindergarten (NCTM 1973, pp. 15–22).

The recommendations for almost all major curriculum groups notwithstanding, close scrutiny of current textbook series, the proposed curriculum for many schools, and classroom observation show that the amount of classroom time given to geometry has decreased in recent years. The following reasons for teaching geometry have been compiled from various sources (NCTM 1970, 1973, Reeve 1930, Wirszup 1976) and are presented to stimulate a renewed interest in geometry as a part of a comprehensive balanced curriculum. The reasons for teaching geometry have been categorized as follows: We teach geometry:

1. to provide for the child's everyday needs
2. to provide for the child's developmental needs
3. to achieve subject matter and content goals
4. to achieve process goals
5. to develop a cultural awareness

Everyday Needs

Certain geometric terms and the ideas they represent are a part of the basic language needed to communicate effectively in daily life. On a trip, the point where two roads *intersect* on a map is discussed. A golf instructor encourages the learner to place the putter *perpendicular* to a line from the cup to the ball. A tennis instructor admonishes budding youngsters to swing with the racquet *parallel* to the ground. A driving test asks the aspirant to distinguish the *shapes* of various road signs. A carpenter's helper is asked to cut a board at a 45-degree *angle*. It is important for persons to learn to communicate with the "language of geometry."

Geometry meets a second everyday need by providing practical, useful procedures. For example, a carpenter uses a procedure for producing a perpendicular bisector of a line segment to find the center of a circular tabletop in order to attach a base. A homeowner uses the concepts of area and volume to accurately determine the quantities of paint, roofing material, grass seed, or lumber to purchase. Whether one is making clothes or following a diagram in a do-it-yourself project, a basic understanding of the ideas, relationships, and procedures in geometry are very helpful. Even in hobbies, such as photography, dome building, or sailing, an understanding of geometry is useful and can enhance the enjoyment.

Beyond this, an understanding of the spatial concepts and relations of geometry is useful to many occupations. As people work in industry, in the building trades, in engineering, in architecture, and in many other professions, geometry can be applied to advantage.

Since many of the things we do involve interaction with objects in a three-dimensional world and since these objects are better understood and manipulated through an understanding of geometry, it is readily apparent that geometry can help meet our everyday needs.

Developmental Needs

A consideration of Piaget's developmental psychology (Flavell 1963), the multiple embodiment principle espoused by Dienes (1964, p. 40), and Bruner's enactive, iconic, and symbolic stages of concept development (Bruner 1966) underscores the fact that many learning theorists support the idea that children's action on objects is a prerequisite to effective learning of concepts. Since objects, even those especially prepared for developing concepts in elementary school mathematics, such as Cuisenaire rods, Dienes multibase blocks, logical blocks, and the like, are essentially geometric in nature, the developmental theories would seem to support the need for a geometric component in the curriculum.

A second major theme that runs throughout Piaget's work is that continuous and progressive changes take place in the structures of behavior and thought in the developing child. Bruner uses this idea in his discussion of a spiral curriculum, and Dienes supports it as he discusses the ever recurring play, structuring, and practice stages in concept development (Dienes 1960, pp. 31–48).

These ideas suggest that concepts are introduced, nurtured, and expanded over a relatively long period of time. If a full-blown concept is achieved in this evolutionary way, it seems that geometric concepts encountered without formal instruction at an early age should be followed by sequential school experiences designed to ever broaden these concepts as children progress through the elementary, junior high, and secondary school.

Recent research has indicated that the two hemispheres of the brain play differing roles in the learning of mathematics (Wheatley et al. 1978, pp. 20–29). It appears that the left hemisphere is involved in the logical, rule-oriented, symbolic, verbal types of thinking. The right hemisphere is involved with spatial, intuitive, nonverbal, gestalt-type thinking. Researchers have recently suggested that for optimum intellectual development, children need opportunities to become involved in more situations that require right-brain processes, and geometry is especially appropriate for this purpose.

In addition to the developmental advantages of geometry suggested above, it should be pointed out that there is evidence that experiences in geometry can do much to help prevent certain learning difficulties as well as help children who encounter learning difficulties. The clinical activities and research described by Frostig (1972) indicate that some of the major visual-perception difficulties can be helped through carefully structured geometric activities. Others, such as Ebersole, Kephart, and Ebersole (1968), have used geometric activities as early experiences to help children over later possible trouble spots in reading. For example, it appears that shape identification activities can help children later discriminate between b's and p's, u's, v's, and so forth.

There are also studies that show a correlation between measures of low spatial ability and measures of mathematics anxiety. In a sense, there are indications that students who have not developed their spatial abilities are more apprehensive and nervous about learning a broad variety of topics in mathematics. This may not be surprising in light of the fact that tasks like reading and interpreting graphs, measuring, evaluating size, and even writing, sequencing, and placing numbers correctly require spatial perception.

Geometry provides other advantages of a psychological nature. For example, anecdotal reports from classrooms in which children encounter geometric ideas indicate that geometric activities are unusually effective in arousing a child's curiosity and motivating the child to become involved. Because geometry is conducive to "hands-on" activities, restless students are stimulated to actively participate. Because of this and the fact that many geometric ideas are simply interesting and fun to deal with, they provide a good change of pace and help develop positive attitudes toward mathematics. For these and other similar reasons, the study of geometry appears advantageous from a psychological and developmental standpoint.

Subject-Matter Goals

Since geometry is a branch of mathematics, the mathematical literacy gained by studying this subject would seem sufficient reason for including it in a comprehensive curriculum. Further, the informal intuitive geometry suggested for kindergarten through grade nine provides a necessary foundation for the logical study of geometry in the secondary school. But if we culminate our discussion here, we miss valuable reasons for introducing geometry in the elementary grades. Some of these reasons are discussed below.

The knowledge of geometric concepts and relationships is vital to a complete understanding of the idea of a number. Manipulative materials of a geometric nature are used to develop basic concepts of number and place value. The number line, a geometric entity, is further used to clarify numbers and their order. At later stages, triangular, rectangular, circular, square, and other regions are used to present the underlying notions of fractional and irrational numbers. Thus, the ideas of geometry provide a basis for readiness, development, extension, and enrichment of the number concept.

Geometric concepts can provide situations for understanding and using algorithms. The geometric concept of area, for example, can be used to provide a rationale for the algorithm for multiplying whole numbers and for multiplying and adding fractional numbers. Geometry also presents problem-solving situations that provide practice in numerical calculations. For example, finding the area of a rectangle provides practice in multiplying; finding the perimeter of a rectangle provides practice in addition. Work with geometric formulas helps children gain facility in combining operations.

Geometric ideas are also prerequisite for the important concept of measurement. An understanding of congruent segments provides a basis for measuring length. A variety of geometric ideas are involved in the concept of measuring area and volume. If children have had preliminary experiences with geometry, they will have developed a readiness for measurement and will find this process much easier to comprehend.

Geometry can also provide a valuable background for activities that involve the presentation and interpretation of data. Line graphs, bar graphs, circle graphs, maps, and the like, all require a facility with geometry for their construction and interpretation.

As one plans a balanced curriculum, ways are often sought to integrate mathematics with other subjects. Geometry can provide a valuable means of accomplishing this. For example, work with ruler and compass designs, construction of solid models, paper-folding activities such as origami, drawing and coloring tessellations of the plane, construction of Escher-type drawings, and many other similar geometric activities can provide extensive integration of geometry with art activities. Also, the geometric concepts of circles, ellipses, angles, spheres, and so on, can be developed and used as students study astronomy and other areas of science. Further, many practical-arts activities involving construction, following patterns, perspective drawing, carpentry, arts and crafts, and so forth, use a host of geometric ideas. Thus, the study of practical arts can motivate the study of geometry and vice versa. Finally, the geometry of geodesics and the geometric notions involved in making and interpreting maps provide avenues for integrating geometry and social science. It is clear that early geometry activities can play an important role in helping meet subject matter goals at all levels.

Process Goals

Bruner (1966) makes these remarks about the importance of process goals:

> Finally, a theory of instruction seeks to take account of the fact that a curriculum reflects not only the nature of knowledge itself (the specific capabilities) but also the nature of the knower and of the knowledge getting process. . . . To instruct someone in these disciplines is not a matter of getting him to commit results to mind.

> Rather it is to teach him to participate in the process that makes this possible—the establishment of knowledge. We teach a subject not to produce little living libraries on that subject but rather to get a student to think mathematically for himself, to conserve matters as a historian does, to take part in the process of knowledge getting. Knowing is a process, not a product. (pp. 59–68)

One process deemed important as the student is "learning how to learn" is closely connected with discovery learning. It is interesting to note that many of the situations conducive to discovery learning are geometric in nature. For example, a child's curiosity can be aroused when considering what type figure is formed by connecting the midpoints of the sides of a quadrilateral. The child is first encouraged to connect the midpoints of the sides of different-sized squares and as a result conjectures that the figure in question is a square (Figure 7-1a). The teacher presents a rhombus and, having connected the midpoints of the different sides of the rhombus, the child tentatively concludes that the resulting figure is at least a rectangle (Figure 7-1b). Finally, the child is encouraged to connect the midpoints of parallelograms, trapezoids, and other quadrilaterals. A refined conjecture is made that when the midpoints of any quadrilateral are connected, the resulting figure is always a parallelogram (Figure 7-1c). Additional examples seem to verify this. Finally, a teacher or another student asks, "What if a four-sided figure has 'dents' in it like the one in Figure 7-1d?" and the conjecture is tested again.

This example illustrates that a geometric setting can be used to help children learn the discovery process. Note that the example above also provides the child with an opportunity to learn the process of formulating and testing conjectures. Later, as the student develops the appropriate background, this situation can again be used to help in learning the process of verification or proof of a generalization.

Many persons who have studied only the rigorous "proof-oriented" geometry of the secondary school find it difficult to appreciate that geometry provides a fertile vehicle for accomplishing other important process goals such as those described as "intuitive thinking" and "divergent thinking."

Polya (1957) suggests asking oneself questions about a problem to help organize thinking about the problem. One of the questions is "Will a picture help?" It has long been recognized that geometry provides an especially important vehicle for modeling or pictorial representation of more abstract situations and is a valuable aid in the problem-solving process.

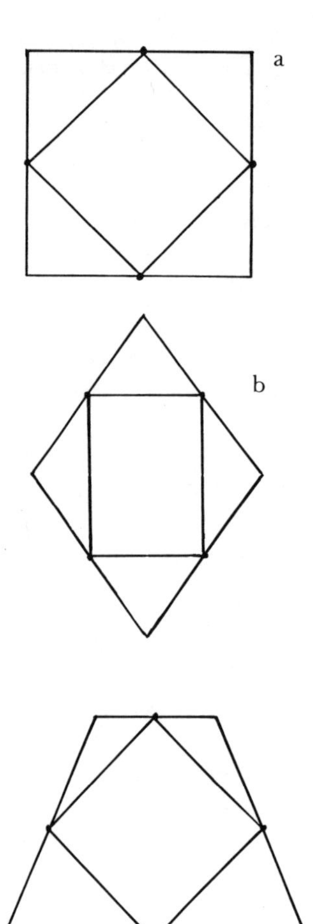

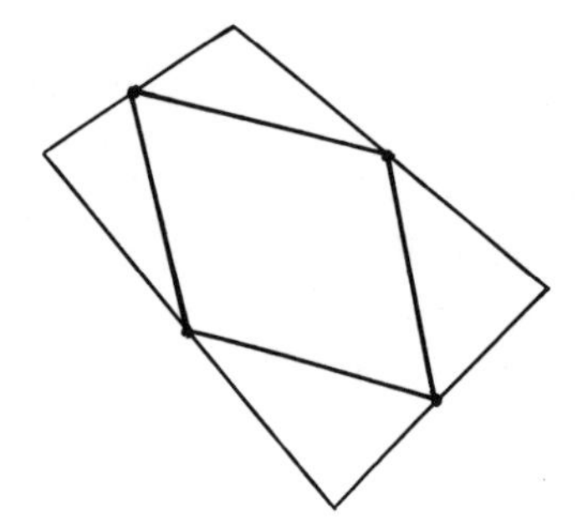

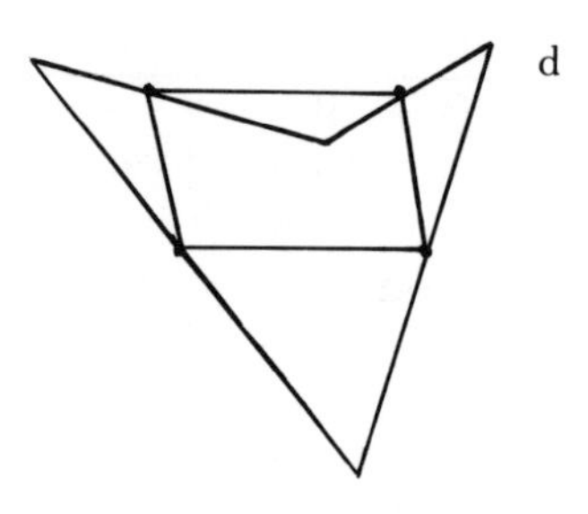

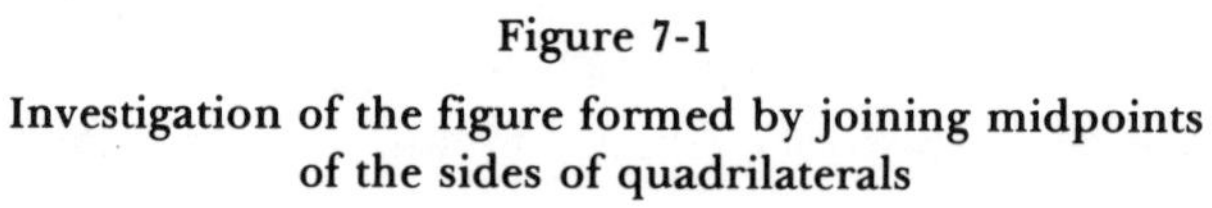

Figure 7-1

Investigation of the figure formed by joining midpoints of the sides of quadrilaterals

If the processes described above, and others, are important goals of instruction, then it seems reasonable to choose content that best provides avenues for developing the processes. Geometry is often the appropriate content.

Cultural Awareness

The importance of geometry as a part of our cultural heritage is underscored by Albert Einstein's comment, refered to earlier in this chapter. Many of the developments in our culture are directly or indirectly related to applications of geometry. If we are to evolve in our appreciation of thought and creative endeavors in a variety of disciplines, it is important that we understand the role that mathematics in general and geometry in particular have played in the development of civilization. In addition, a foundation in geometry provides a basis for art appreciation, an appreciation of architecture, and a foundation for appreciating many of the geometric recreations that have captivated persons in all walks of life for centuries.

THE ROLE OF GEOMETRY IN A COMPREHENSIVE, BALANCED CURRICULUM

The rationales offered in the preceding section not only serve to justify teaching geometry in the elementary, junior high, and secondary schools, but they also influence the selection and sequencing of geometric ideas in the curriculum. While it is beyond the scope of this chapter to describe completely the role of geometry in the curriculum for kindergarten through grade twelve, the following statements are presented to encourage teachers and curriculum developers to analyze their positions on this issue. It is hoped that these statements will be tested against experience in the classroom, theories of teaching and learning geometry, and the reasons given for teaching geometry in the schools.

1. Geometry should be a major strand in the curriculum for kindergarten through grade twelve. Consideration should be given to everyday needs, developmental needs, subject matter goals, process goals, and the development of cultural awareness.
2. Geometric activities at all levels should be central parts of sequentially planned lessons based on specific objectives rather than presented as supplementary or enrichment exercises. (Attention should be given to development of concepts, discovery

of generalizations, development of basic construction and other skills, and the development of higher-level skills such as proof and problem solving.

3. Early experiences with geometry should include extensive work with materials such as logical blocks, geoboards, geostrips, pattern blocks, plastic or cardboard models and cutouts, tangrams, mirrors, mirror cards, tessellation pieces, and so forth. The focus should be on intuitive, physically oriented experiences designed to develop geometric concepts. Emphasis on symbolism and formal definition should be delayed until the children have had ample opportunity for active involvement with geometry at the physical, informal level.
4. Subsequent work in the middle grades should build upon the early physical experiences by continuing to use manipulative materials to explore the important relationships of geometry such as parallel, perpendicular, congruent, and similar. The geometry of motions and symmetry should also be explored. Geometric constructions using the Mira as well as the ruler and compass should be based upon clarifying and extending concepts developed earlier, leading toward the use of some symbols and simple definitions to describe the basic ideas. The relational concepts should be also emphasized.
5. At the junior high level, extensive use should be made of paper folding, model building, geometric constructions, and measurement of segments and angles to aid the students in discovering generalizations. Simple problem solving and formulas should be included.
6. At the secondary level, additional intuitive experiences should be provided to review and extend informal ideas. Definitions should be emphasized and used with postulates and logic to prove theorems. An appropriate balance between inductive and deductive experiences should be effected. The use of geometry in problem-solving applications should be emphasized.

After the basic concepts of geometry are understood and measurement and construction skills have been developed, much of the work in geometry involves seeking, proving, and using generalizations. Figure 7-2 suggests a sequence of events for dealing with generalizations in geometry.

This particular sequence of events, sometimes used to suggest a lesson plan for teaching geometry at the secondary level, contains some elements that deserve the attention of those involved in curriculum development. First, it emphasizes the potential role that the discovery of generalizations could play in learning geometry. Many people have had geometry experiences that have been restricted to the proof of the theorems that have been given them, yet it is well known that inductive reasoning plays a major role in the development of mathematics. It has also been established that junior high and secondary level students, with reasonable aid from the text and teacher, can discover many of the key generalizations. Unless they are given opportunities to do this, they miss an important aspect of this subject.

Second, the sequence suggests that for both motivation and application, problem solving in a geometric setting is important. Further, one might interpret the reference to "applied problems" as indicating that the uses of geometry in the real world should be emphasized. Finally, one could observe that while proving generalizations (theorems) is a major goal at the secondary level, it would be inappropriate to focus exclusively on this goal at the expense of providing opportunities for discovery and applications.

Thus the comprehensive, balanced curriculum includes a sequential development of geometry introduced with physical materials, continually related to the real world, and designed to provide opportunities for discovery, proof, and problem solving. Both content and process goals are emphasized, and one of the expected outcomes is that children will enjoy this study.

Motivation — (A problem or puzzle situation that entices exploration of the generalization)

↓

Inductive Reasoning — (Discovery of the generalization)

↓

Deductive Reasoning — (Verification or proof of the generalization)

↓

Application — (Using the generalization to solve applied problems)

Figure 7-2

Sequence of events involving generalizations

SOME FINAL QUESTIONS

In conclusion, a summary of some of the issues facing those who are currently making decisions about the role of geometry in the schools will be highlighted by the following questions:

1. How should the geometry program at each level be best sequenced and presented so as to have coherence, continuity, and purpose?
2. How can teachers and parents be informed and convinced that geometry is an important subject that should be considered an integral part of the curriculum at all levels?
3. How can enough time be found at the elementary level to include a significant amount of geometry and still adequately teach the basic ideas of number and computation?
4. How can a geometry program be planned that requires manipulative materials and assurance be provided that the materials will be available in the schools and teachers will have the necessary expertise to use them?
5. Which activities of geometry can be effectively presented to which students at which level?
6. What should be the balance between content and process goals at every level?
7. What should be the balance between induction and deduction (discovery versus proof) at every level?
8. What approach to geometry is most appropriate, especially at the secondary level: standard approach, transformation approach, coordinate geometry approach, vector approach, another approach, or a combination of these? (See NCTM 1973 for further reading on these issues.)
9. Should logic be taught? If so, at what levels? Is it more effectively taught in the context of geometry or as "logic in everyday life?"

Perhaps the reader can add to this list of "issues" questions and will find that the process of seeking additional answers will aid in the curriculum improvement process. Also it is hoped that future research in psychology, education, and mathematics education will help provide more definitive answers and a more secure framework upon which to make curricular decisions.

The simple message of this chapter is that geometry is a significant, worthwhile branch of knowledge and that there are some pervasive reasons for teaching it at all levels. While there is currently no complete agreement on questions of what, where, and how, many of the issues have been defined, and some guiding principles are available for consideration. A balanced, comprehensive curriculum can only be achieved if principles such as these are carefully evaluated, and decisions based on this evaluation are implemented in the classroom.

REFERENCES

Bruner, Jerome S. *Toward a Theory of Instruction.* Cambridge, Mass.: Belknap Press, 1966.

Dienes, Zoltan P. *Building up Mathematics.* London: Hutchinson Educational, 1960.

Dienes, Zoltan P. *The Power of Mathematics.* London: Hutchinson Educational, 1964.

Ebersole, Marylou; Kephart, Newell C.; and Ebersole, James B. *Steps to Achievement for the Slow Learner.* Columbus, Ohio: Charles E. Merrill, 1968.

Flavell, John H. *The Developmental Psychology of Jean Piaget.* Princeton, N.J.: Van Nostrand Co., 1963.

Frostig, Marianne. *Pictures and Patterns.* Teacher's Guides. Chicago: Follett Publishing Co., 1972.

National Council of Teachers of Mathematics. *Readings in Geometry from the Arithmetic Teacher*, edited by Marguerite Brydegaard and James E. Inskeep, Jr. Washington, D.C.: NCTM, 1970.

National Council of Teachers of Mathematics. *Geometry in the Mathematics Curriculum.* Thirty-sixth Yearbook. Reston, Va.: NCTM, 1973.

O'Daffer, Phares G., and Clemens, Stanley R. *Geometry: An Investigative Approach.* Boston: Addison-Wesley, 1976.

Polya, George. *How to Solve It.* 2d ed. Garden City, N.Y.: Doubleday, 1957.

Reeve, William D. "The Teaching of Geometry." In *The Teaching of Geometry.* Fifth Yearbook of the National Council of Teachers of Mathematics. New York: Bureau of Publications, Teachers College, Columbia University, 1930, pp. 1–28.

Wheatley, Grayson H.; Mitchell, Robert; Frankland, Robert L.; and Kraft, Rosemarie. "Hemispheric Specialization and Cognitive Development: Implications for Mathematics Education." *Journal for Research in Mathematics Education* 9 (January 1978): 20–29.

Wirszup, Izaak. "Breakthroughs in the Psychology of Learning and Teaching Geometry." In *Space and Geometry*, edited by J. Larry Martin. Columbus, Ohio: ERIC Information Analysis Center for Science, Mathematics, and Environmental Education, 1976, pp. 76–79.

PART THREE
Instructional Aids

8. The Role of Manipulative Materials in the Learning of Mathematical Concepts

Thomas R. Post

It should not be surprising that current research has established a substantial relationship between the use of manipulative materials and students' achievement in the mathematics classroom. Learning theorists have suggested for some time that childrens' concepts evolve through direct interaction with the environment, and materials provide a vehicle through which this can happen. This message has been conveyed in a number of ways: Piaget (1971) suggested that concepts are formed by children through a reconstruction of reality, not through an imitation of it; Dewey (1938) argued for the provision of firsthand experiences in a child's educational program; Bruner (1960) indicated that knowing is a process, not a product; and Dienes (1969), whose work specifically relates to mathematics instruction, suggested that children need to build or construct their own concepts from within rather than having those concepts imposed upon them.

Researchers in mathematics education are in the process of accumulating a persuasive body of evidence that supports the use of manipulative materials in the mathematics classroom. In view of this, it is perplexing that relatively few programs incorporate a substantive experimental component while so many others concentrate merely on completing the pages in the ubiquitous commercially produced textbooks and workbooks. This chapter will discuss the theoretical rationale for using manipulative materials in the classroom, provide a summary of what is already known about the impact of manipulative materials on mathematics learning, discuss barriers to their use, and suggest directions for their use in future research.

THE NATURE OF THE PROBLEM OF LEARNING MATHEMATICS

The nature of mathematics is independent of us personally and of the world outside." (Jourdain)

Mathematics in its purest sense is an abstraction. Whether it was discovered by or has been created by mankind is perhaps a philosophical point and need not concern us here; but the fact is that it exists, and it is extremely useful in describing and predicting events in the world around us. How then is it so useful if it exists "independent of us personally and of the world outside?" The answer lies in the ability of mathematics to model effectively numerous aspects of the real world. It does this by creating abstract structures that have properties or attributes similar to its real-world counterpart. If the model behaves in a manner that truly parallels the original, then it becomes possible to manipulate and use the model to make conclusions and/or predictions about its counterpart in the real world. We can do this because we know the two systems "behave" in the same manner and because we know that an operation in one system will have its counterpart in the other. This can be depicted as shown in Figure 8-1.

Lesh (1979) has suggested that manipulative materials can be effectively used as an intermediary between the real world and the mathematical world. He contends that such use would tend to promote problem-solving ability by providing a vehicle through which children can model real-world situations. The use of manipulative materials (concrete models) in this manner is thought to be more abstract than the actual situation yet less abstract than the formal symbols. Figure 8-2 illustrates the revised model. It should be noted that this expanded use departs from the more traditional classroom technique wherein manipulatives have been used to teach children how to calculate using the four arithmetic operations.

Relying on the model depicted in Figure 8-2, we find it possible to determine the total number of milk containers needed by the three first-grade classes in Main Elementary School by adding the numbers

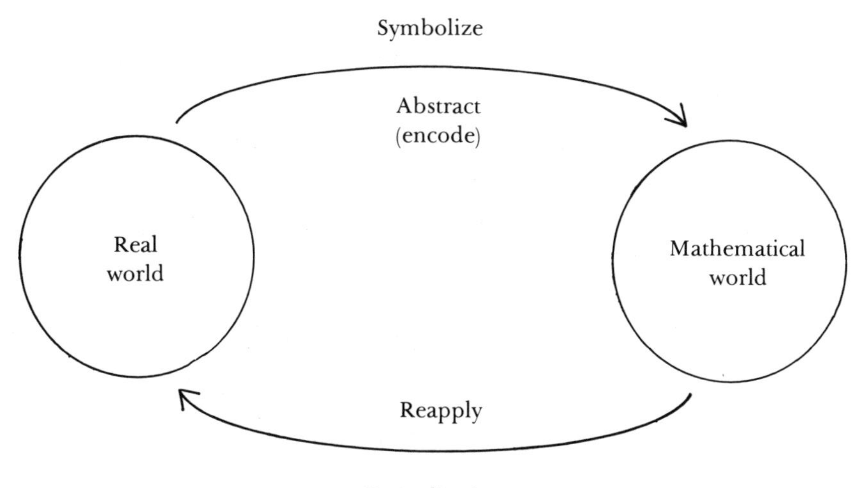

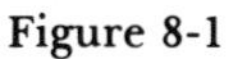
Figure 8-1

A relationship between the real and mathematical worlds

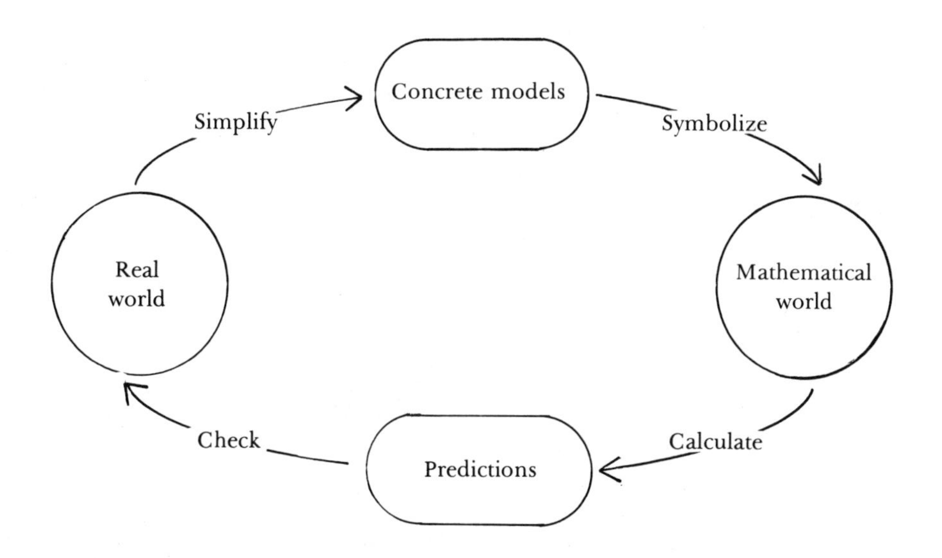

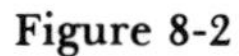
Figure 8-2

An extended relationship between the real and mathematical worlds

twenty-four, twenty-seven, and twenty-five. It is not necessary to pair each child with a milk container in order to find the correct total because the abstract system of addition is structurally similar to the problem in question and therefore can be used as a model for it. More important, this abstract system is structurally similar to all physical situations where a sum corresponding to the union of a number of disjoint sets is required. It now becomes possible to utilize this more abstract and admittedly less cumbersome system to make conclusions about the more concrete and awkward system.

Recall that the problem originated in the real world and was concerned with identifying a one-to-one correspondence between the number of individuals and a like number of milk containers. The problem was then changed into a more suitable, though admittedly more abstract, format; and from a practical standpoint, a considerably more manageable format. It is important to note that this transformation of the problem situation preserved all the important structural aspects of that situation. To complete the problem, the numbers are added, a sum is generated, and conclusions are related to the real-world situation at Main Elementary School. We have come full cycle (see Figure 8-2).

The structural similarity between these two systems is known as an isomorphism. It is an extremely important concept in mathematics, for if any two systems can be shown to be isomorphic to one another, it becomes possible to work in the simpler or more available system and transfer all conclusions to the less accessible one. In reality, complete isomorphisms are never really established between an abstract concept and a set of physical materials or a real-world situation. The extent to which the partial isomorphism approximates the concept is the extent to which the more accessible structure is useful in teaching the concept. The fact that some sets of manipulative materials are better than others for teaching a particular concept attests to this.

Number is an abstraction. No one has ever seen a number and no one ever will. "Twoness" is an idea. We see illustrations of this idea everywhere, but we do not see the idea itself. In a similar way the symbol "2" is used to elicit a whole series of recollections and experiences that we have had entailing the concept of two, but the squiggly line *2* in and of itself is not the concept.

How then do we teach children about the concept of number if as indicated it is a total abstraction? The answer is very much related to the concept of an isomorphism. For if a parallel structure that was more accessible and perhaps manipulable could be identified having the same properties as the set of whole numbers, then it would be possible to operate within this more accessible (and isomorphic) structure and subsequently to make conclusions about the more abstract system of number. This is precisely what happens. Sometimes these artificially constructed systems are called interpretations or embodiments of a concept. Some examples of these partial isomorphisms are using counters or sets of objects to represent the counting numbers (a discrete model), using lengths such as number lines or Cuisenaire rods to represent the set of real numbers (a continuous model), and using the area of a rectangle to represent the multiplication of two whole numbers or fractions. Manipulative materials may now be viewed simply as isomorphic structures that represent the more abstract mathematical notions we wish to have children learn.

SEVERAL THEORETICAL PERSPECTIVES

The major theoretical rationale for the use of manipulative materials in a laboratory-type setting has been attributed to the works of Piaget, Bruner, and Dienes. Each represents the cognitive viewpoint of learning, a position that differs substantially from the connectionist theories that were predominant in educational psychology during the first part of the twentieth century. Modern cognitive psychology places great emphasis on the process dimension of the learning process and is at least as concerned with "how" children learn as with "what" it is they learn. The objective of true understanding is given highest priority in the teaching/learning process, and it is generally felt that such understanding can only follow the individuals' personalized perception, synthesis, and assimilation of relationships as these are encountered in real situations. Emphasis is placed, therefore, on the interrelationships between parts as well as the relationship between parts and whole.

Each of these men subscribes to a basic tenet of gestalt psychology, namely that the whole is greater than the sum of its parts. Each suggests that the learning of large conceptual structures is more important than the mastery of large collections of isolated bits of information. Learning is thought to be intrinsic and, therefore, intensely personal in nature. It is the meaning that each individual attaches to an experience which is important. It is generally felt that the degree of meaning is maximized when individuals are allowed and encouraged to interact personally with various aspects of their environment. This, of course, includes other people. It is the physical action on the part of the child that contributes to her or his understanding of the ideas encountered.

Proper use of manipulative materials could be used to promote the broad goals alluded to above. I will discuss each of these men more fully since each has made distinct contributions to a coherent rationale for the use of manipulative materials in the learning of mathematical concepts.

Jean Piaget

Piaget's contributions to the psychology of intelligence have often been compared to Freud's contributions to the psychology of human personality. Piaget has provided numerous insights into the development of human intelligence, ranging from the random responses of the young infant to the highly complex mental operations inherent in adult abstract reasoning. He has established the framework within which a vast amount of research has been conducted, particularly within the past two decades.

In his book *The Psychology of Intelligence* (1971), Piaget formally develops the stages of intellectual development and the way they are related to the development of cognitive structures. His theory of intellectual development views intelligence as an evolving phenomenon occurring in identifiable stages having a constant order. The age at which children attain and progress through these stages is variable and depends on factors such as physiological maturation, the degree of meaningful social or educational transmission, and the nature and degree of relevant intellectual and psychological experiences.

Piaget regards intelligence as effective adaptation to one's environment. The evolution of intelligence involves the continuous organization and reorganization of one's perceptions of, and reactions to, the world around him. This involves the complementary processes of assimilation (fitting new situations into existing psychological frameworks) and accommodation (modification of behavior by developing or evolving new cognitive structures). The effective use of the assimilation-accommodation cycle continually restores equilibrium to an individual's cognitive framework. Thus the development of intelligence is viewed by Piaget as a dynamic, nonstatic evolution of newer and more complex mental structures.

Piaget's now-famous four stages of intellectual development (sensorimotor, preoperational, concrete operations, and formal operations) are useful to educators because they emphasize the fact that children's modes of thought, language, and action differ both in quantity and quality from that of the adult. Piaget has argued persuasively that children are not little adults and therefore cannot be treated as such.

"Perhaps the most important single proposition that the educator can derive from Piaget's work, and its use in the classroom, is that children, especially young ones, learn best from concrete activities" (Ginsberg and Opper 1969, p. 221). This proposition, if followed to its logical conclusion, would substantially alter the role of the teacher from expositor to one of facilitator, that is, one who promotes and guides children's manipulation of and interaction with various aspects of their environment. While it is true that when children reach adolescence their need for concrete experiences is somewhat reduced because of the evolution of new and more sophisticated intellectual schemas, it is not true that this dependence is eliminated. The kinds of thought processes so characteristic of the stage of concrete operations are in fact utilized at all developmental levels beyond the ages of seven or eight. Piaget's crucial point, which is sometimes forgotten or overlooked, is that until about the age of eleven or twelve, concrete operations represent the highest level at which the child can effectively and consistently operate. Piaget has emphasized the important role that social interaction plays in both the rate and quality with which intelligence develops. The opportunity to exchange, discuss, and evaluate one's own ideas and the ideas of others encourages decentration (the diminution of egocentricity), thereby leading to a more critical and realistic view of self and others.

It would be impossible to incorporate the essence of these ideas into a mathematics program that relies primarily (or exclusively) on the printed page for its direction and "activities." To be sure, Piaget speaks to much more than the learning of mathematics per se. Intellectual development is inextricably intertwined with the social/psychological development of children, but it should be noted that mathematics and science, with their wide diversity of ideas and concepts and their capacity for being represented by concrete isomorphic structures, are especially well suited to the promotion of these ends.

It is generally felt that the basic components of a theoretical justification for the provision of active learning experiences in the mathematics classroom are embedded in Piaget's theory of cognitive development. Dienes and Bruner, while generally espousing the views of Piaget, have made contributions to the cognitive view of mathematics learning that are distinctly their own. The work of these two men lends additional support to this point of view.

Zoltan P. Dienes

Unlike Piaget, Dienes has concerned himself exclusively with mathematics learning; yet like Piaget, his major message is concerned with providing a justification for active student involvement in the learning process. Such involvement routinely involves the use of a vast amount of concrete material.

Rejecting the position that mathematics is to be learned primarily for utilitarian or materialistic reasons, Dienes (1969) sees mathematics as an art form to be studied for the intrinsic value of the subject itself. He believes that learning mathematics should ultimately be integrated into one's personality and thereby become a means of genuine personal fulfillment. Dienes has expressed concern with many aspects of the status quo, including the restricted nature of mathematical content considered, the narrow focus of program objectives, the overuse of large-group instruction, the debilitating nature of the punishment-reward system (grading), and the limited dimension of the instructional methodology used in most classrooms.

Dienes's theory of mathematics learning has four basic components or principles. Each will be discussed briefly and its implications noted. (The reader will notice large-scale similarities to the work of Piaget.)

The Dynamic Principle. This principle suggests that true understanding of a new concept is an evolutionary process involving the learner in three temporally ordered stages. The first stage is the preliminary or play stage, and it involves the learner with the concept in a relatively unstructured but not random manner. For example, when children are exposed to a new type of manipulative material, they characteristically 'play' with their newfound 'toy.' Dienes suggests that such informal activity is a natural and important part of the learning process and should therefore be provided for by the classroom teacher. Following the informal exposure afforded by the play stage, more structured activities are appropriate, and this is the second stage. It is here that the child is given experiences that are structurally similar (isomorphic) to the concepts to be learned. The third stage is characterized by the emergence of the mathematical concept with ample provision for re-application to the real world. This cyclical pattern can be depicted as shown in Figure 8-3.

The completion of this cycle is necessary before any mathematical concept becomes operational for the learner. In subsequent work Dienes elaborated upon this process and referred to it as a learning cycle (Dienes 1971, Dienes and Golding 1971). The dynamic principle establishes a general framework within which learning of mathematics can occur. The remaining components should be considered as existing within this framework.

The Perceptual Variability Principle. This principle suggests that conceptual learning is maximized when children are exposed to a concept through a variety of physical contexts or embodiments. The experiences provided should differ in outward appearance while retaining the same basic conceptual structure. The provision of multiple experiences (not the same experience many times), using a variety of materials, is designed to promote *abstraction* of the mathematical concept. When a child is given opportunities to see a concept in different ways and under different conditions, he or she is more likely to perceive that concept irrespective of its concrete embodiment. For example, the regrouping procedures used in the process of adding two numbers is independent of the type of materials used. We could therefore use tongue depressors, chips, and abacus, or multibase arithmetic blocks to illustrate this process. When exposed to a number of seemingly different tasks that are identical in structure, children will

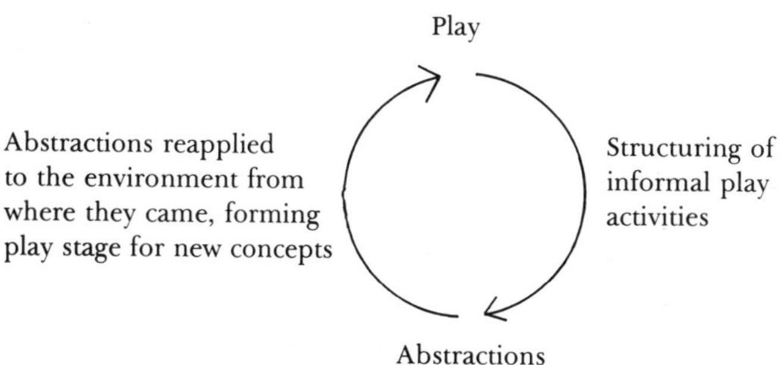

Figure 8-3

Dienes's learning cycle

tend to abstract the similar elements from their experiences. It is not the performance of any one of the individual tasks that is the mathematical abstraction but the ultimate realization of their similarity.

The Mathematical Variability Principle. This principle suggests that the *generalization* of a mathematical concept is enhanced when the concept is perceived under conditions wherein variables irrelevant to that concept are systematically varied while keeping the relevant variables constant. For example, if one is interested in promoting an understanding of the parallelogram, this principle suggests that it is desirable to vary as many of the irrelevant attributes as possible. In this example, the size of angles, the length of sides, the position on the paper should be varied while keeping the relevant attribute—opposite sides parallel—intact. Dienes suggests that the two variability principles be used in concert with one another since they are designed to promote the complementary processes of abstraction and generalization, both of which are crucial aspects of conceptual development.

The Constructivity Principle. Dienes identifies two kinds of thinkers: the constructive thinker and the analytic thinker. He roughly equates the constructive thinker with Piaget's concrete operational stage and the analytical thinker with Piaget's formal operational stage of cognitive development.

This principle states simply that "construction should always precede analysis." It is analogous to the assertion that children should be allowed to develop their concepts in a global intuitive manner emanating from their own experiences. According to Dienes, these experiences carefully selected by the teacher form the cornerstone upon which all mathematics learning is based. At some future time, attention will be directed toward the analysis of what has been constructed; however, Dienes points out that it is not possible to analyze what is not yet there in some concrete form.

Summary and Implications. The unifying theme of these four principles is undoubtedly that of stressing the importance of learning mathematics by means of direct interaction with the environment. Dienes is continually implying that mathematics learning is not a spectator sport and, as such, requires a very active type of physical and mental involvement on the part of the learner. In addition to stressing the environmental role in effective conceptual learning, Dienes addresses, in his two variability principles, the problem of providing for individualized learning rates and learning styles. His constructivity principle aligns itself closely with the work of Piaget and suggests a developmental approach to the learning of mathematics that is temporally ordered to coincide with the various stages of intellectual development. The following are some implications of Dienes's work.

1. The *class* lesson would be greatly deemphasized in order to accommodate individual differences in ability and interests.
2. Individual and small-group activities would be used concomitantly, since it is not likely that more than two to four children would be ready for the same experience at the same time.
3. The role of teacher would be changed from expositor to facilitator.
4. The role of students would be expanded to include the assumption of a greater degree of responsibility for their own education.
5. The newly defined learning environment would create new demands for additional sources of information and direction. The creation of a learning laboratory containing a large assortment of both hardware and software would be a natural result of serious consideration of Dienes's ideas (Reys and Post 1973).

Jerome Bruner

Greatly influenced by the work of Piaget and having worked for some time with Dienes at Harvard, Bruner shares many of their views. Interested in the general nature of cognition (conceptual development), he has provided additional evidence suggesting the need for firsthand student interaction with the environment. His widely quoted (and hotly debated) view that "any subject can be taught effectively in some intellectually honest form to any child at any stage of development" (Bruner 1966, p. 33) has encouraged curriculum developers in some disciplines (especially social studies) to explore new avenues of both content and method. In recent years Bruner has become widely known in the field of curriculum development through his controversial elementary social studies program, *Man: A Course of Study* (1969).

Bruner's instructional model is based on four key concepts: structure, readiness, intuition, and motivation. These constructs are developed in detail in his classic book, *The Process of Education* (1960).

Bruner suggests that teaching students the structure of a discipline as they study particular content leads to greater active involvement on their part as they discover basic principles for themselves. This, of course, is very different from the learning model that suggests students be receivers rather than developers of information. Bruner states that learning the structure of knowledge facilitates comprehension, memory, and transfer of learning. The idea of structure in learning leads naturally to the process approach where the very process of learning (or *how* one learns) becomes as important as the content of learning (or *what* one learns). This position, misunderstood by many, has been the focus of considerable controversy. The important thing to remember is that Bruner never says that content is unimportant.

Bruner (1966) suggests three modes of representational thought. That is, an individual can think about a particular idea or concept at three different levels. "Enactive" learning involves hands-on or direct experience. The strength of enactive learning is its sense of immediacy. The mode of learning Bruner terms "iconic" is one based on the use of the visual medium: films, pictures, and the like. "Symbolic" learning is that stage where one uses abstract symbols to represent reality.

Bruner feels that a key to readiness for learning is intellectual development, or how a child views the world. Here he refers to the work of Piaget, stating that "what is most important for teaching basic concepts is that the child be helped to pass progressively from concrete thinking to the utilization of more conceptually adequate modes of thought" (Bruner 1960, p. 38).

Bruner suggests that readiness depends more upon an effective mix of these three learning modes than upon waiting until some imagined time when children are capable of learning certain ideas. Throughout his writing is the notion that the key to readiness is a rich and meaningful learning environment coupled with an exciting teacher who involves children in learning as a process that creates its own excitement. Bruner clings to the idea of intrinsic motivation—learning as its own reward. It is a refreshing thought.

In short, a general overhaul of existing pedagogical practices, teacher-pupil interaction patterns, mathematical content, and mode of presentation, as well as general aspects of classroom climate would be called for if the views of Piaget, Dienes, and Bruner were to be taken seriously. Each in his own way would promote a revolution in school curricula, one whose major focus would be method as well as content.

CURRENT RESEARCH ON USE OF MANIPULATIVES

Research dealing with the impact of activity-based approaches on the teaching and learning of mathematics is relatively extensive. No less than twenty reviews of research, surveys of the state of the art, or historical overviews have been completed since 1957.

Given the sheer number of studies undertaken, it is perplexing to note that more is not known about the precise way in which manipulative materials affect the development of mathematical concepts. Perhaps the largest contributing factor to this has been the lack of coordinated research efforts that have mapped out a priori an area or areas of investigation and have designed individual investigations that would have provided coordinated answers to sets of related questions. Rather, the past pattern of research has been that of large numbers of individually conducted investigations and then posteriori attempts to relate them in some fashion. This has not been particularly fruitful and has left many unanswered questions and huge gaps in our knowledge.

The most recent and comprehensive review of research on the use of manipulative materials was compiled at the Mathematics and Science Information Reference Center at Ohio State University (ERIC) by Suydam and Higgins (1976). The report generally concludes that manipulatives are effective in promoting student achievement but emphasizes the need for additional research. The impact of the use of manipulative materials upon achievement in mathematics is summarized in Table 8-1 (Suydam and Higgins 1976, pp. 33–39).

The reader will note from Table 8-1 that 60 percent of the research studies examined favored the manipulative treatments, while only 10 percent clearly favored the nonmanipulative treatment. If studies in which no significant differences were found are interpreted as efforts

Table 8-1

Summary of grade-related studies dealing with the impact of manipulative materials on students' achievement

Grade level	Number of studies favoring manipulative materials	Number of studies favoring nonmanipulative materials	Number of studies showing no significant differences	Total
1 and 2	7	2	3	12
3 and 4	9	1	3	13
5 and 6	6	0	3	9
7 and 8	2	1	3	6
	24	4	12	40

Sixteen other studies examined in this report did not fall neatly into one of these three categories.

that did not inhibit achievement, then in 90 percent of the studies reviewed the use of manipulative materials produced equivalent or superior student performance when compared with nonmanipulative approaches. This has led Suydam and Higgins to conclude that

> . . . across a variety of mathematical topics, studies at every grade level support the importance of the use of manipulative materials. Additional studies support the use of both materials and pictures. We can find little conclusive evidence that manipulative materials are effective only at lower grade levels. The use of an activity approach involving manipulative materials appears to be of importance for all levels of the elementary school (1976, p. 60).

The results summarized in the Suydam and Higgins report are similar in nature to those found in earlier reviews of research dealing with manipulative materials and/or mathematics laboratories (Fennema 1972, Fitzgerald 1972, Kieren 1969, 1971, Vance and Kieren 1971, Wilkinson 1974). It was observed that "In almost all cases there is a similarity between their conclusions and certain of ours" (Suydam and Higgins 1976, p. 85).

In conclusion Suydam and Higgins state, "We believe that lessons involving manipulative materials will produce greater mathematical achievement than will lessons in which manipulative materials are not used *if the manipulative materials are used well*" (p. 92). What does it mean to use materials well? The following suggestions were made by Suydam and Higgins (1976, pp. 92–94):

1. Manipulative materials should be used frequently in a total mathematics program in a way consistent with the goals of the program.
2. Manipulative materials should be used in conjunction with other aids, including pictures, diagrams, textbooks, films, and similar materials.
3. Manipulative materials should be used in ways appropriate to mathematics content, and mathematics content should be adjusted to capitalize on manipulative approaches.
4. Manipulative materials should be used in conjunction with exploratory and inductive approaches.
5. The simplest possible materials should be employed.
6. Manipulative materials should be used with programs that encourage results to be recorded symbolically.

Other aspects of manipulative materials need to be considered and researched. Bruner's three modes of representational thought suggest a linear sequence for advancing a particular concept from the concrete to the abstract level: first enactive, then iconic, then symbolic.

Since manipulative materials are a means to an end and not an end in themselves, the intellectual mechanisms used in the transition from one mode to another are of great interest to the researcher and of equal significance to the classroom teacher. At this time, the general nature of those mechanisms is not known. It is known, however, that experience and understanding at one level do not necessarily imply the ability to function at a more sophisticated level. For this reason the translation processes both within modes (multiple embodiments) and between modes need conscious attention.

Under normal conditions, children are given materials during an instructional sequence wherein concepts are introduced and developed. In general, insufficient attention is paid to the way in which this enactive experience is related to the symbolic representation of that experience. For example, children might manipulate blocks, an abacus, or tongue depressors during initial exposure to the concept of place value. They may, however, never receive instruction as to how these materials reflect, in a concrete manner, the abstract manipulation of symbols that is sure to follow the enactive experience; that is, the existing isomorphism between different modes of representation of an idea is never consciously established. When children are then evaluated to assess their levels of achievement, they are expected to perform not at the enactive level, wherein the concept has been introduced and developed, but at the symbolic level. This inconsistency between mode of instruction and mode of evaluation has no doubt resulted in many spurious (and probably negative) conclusions regarding the nature of the impact of manipulative materials upon conceptual development.

One way in which the relationships between enactive and symbolic modes can be highlighted is to juxtapose them, gradually fading out the more concrete mode. That is:

Enactive(E) → Enactive/Symbolic(ES) → Symbolic/Enactive(SE) → Symbolic(S)

This model suggests that a concept should be introduced enactively, with the initial emphasis solely on physical manipulation (E). Next, although the primary emphasis is still on physical manipulation, the child is asked to simply *record* the results of his activity (ES). Third, the child is asked to perform the manipulation symbolically and to *check or reaffirm* symbolic results by reenacting or modeling the original problem or exercise using the manipulative materials (SE). Last, the materials are faded out altogether, and the child operates exclusively at the symbolic level (S). Success within this last phase would seem to be logically dependent upon previous experiences within the other three. Note that the iconic phase is not represented. It is, at present, unclear how to best utilize this mode in the instructional sequence, and yet it would seem to be a valuable adjunct to concrete experience. At present, however, research is inconclusive on this point (Suydam and Higgins 1976, pp. 24–25).

BARRIERS TO THE USE OF MANIPULATIVES

Why are manipulative materials not more extensively used, given the persuasiveness of the theoretical arguments for providing enactive experiences in the mathematics classroom? Classroom teachers, when responding to this question, initially suggest that lack of financial resources is the most important factor inhibiting more extensive use of manipulative materials. The actual reasons are undoubtedly more complex than this.

It is suggested here that inertia and subtle but powerful inschool political pressures are the two most significant factors retarding movement toward expanded use of enactive experiences in the nation's schools. There is evidence that concrete experiences are used in elementary schools. However, one survey indicated that as of 1978, 9 percent of the nation's mathematics classes (kindergarten through grade six) never use materials, and 37 percent use them less than one time per week (Fey 1979, p. 12).

The fact is that systematic use of materials is more difficult for the teacher than administering a program designed around texts and workbooks. School-age children are (or should be) in the process of learning to be responsible for their own actions and learning to control their own behavior. It is, therefore, difficult for them to complete large interrupted segments of on-task time, especially when that time appears on the surface to be more loosely structured than the systematic completion of textbook pages. As a result of the inevitable (but temporary) transitional problems related to classroom management and control, many teachers have given up on such laboratory approaches prematurely, that is, before students have had adequate opportunity to develop the necessary degree of self-control. It must be noted that such decisions are made for reasons related to management and control and not necessarily for reasons related to pedagogy or learning.

These are separate (although related) issues and should not be confused, which unfortunately they often are.

The accountability issue also serves to inhibit widespread departures from the status quo. When accomplishment is viewed in terms of "covering" pages in the textbook, use of extra text activities seems antithetical and counterproductive. Further, when success in the overall program is determined by the extent to which students are able to calculate at the symbolic level on some standardized instrument, widespread use of manipulative materials seems almost counterproductive. Until the public realizes that a test score cannot be interpreted as a valid instrument of true understanding and that the things thus measured may not, in the final analysis, be the most significant outcomes of the mathematics program, this situation is likely to remain unaltered.

The nature of basic skills and learning in mathematics is defined quite differently by the lay population, by classroom teachers, and by university-level mathematics educators. As an expanded definition of the basic skills similar to that suggested by the National Council of Supervisors of Mathematics (NCSM 1978) becomes more widely accepted by the education and lay communities, the nature of the outcomes that classroom teachers are held accountable for will be likewise expanded. This should have the positive effect of expanding the use of enactive experiences as a learning method. This, in turn, will affect the degree to which manipulatives are considered an important aspect of a mathematics program. The current "back-to-basics" movement is fundamentally inconsistent with such an expanded view of the nature and scope of the basic skills in mathematics.

Jackson (1979, pp. 76–78) identified common mistaken beliefs and subsequent abuses resulting from an overzealous acceptance of manipulative materials as the long-sought-after educational panacea. The following were included in his list of mistaken beliefs.

1. Almost any manipulative aid may be used to teach any given concept.
2. Manipulative aids necessarily simplify the learning of mathematical concepts.
3. Good mathematics teaching always accompanies the use of manipulative materials.
4. The more manipulative aids used for a single concept, the better the concept is learned.
5. A single multipurpose aid should be used to teach all or most mathematical concepts.

6. Manipulative aids are more useful in the primary grades than in the intermediate and secondary grades, more useful with low-ability students than with high-ability students.

The matter of whether or not to use manipulative materials in the mathematics classroom is a multifaceted one. It seems clear that in the daily routines of the average classroom, the dilemmas surrounding the use of manipulative aids are complicated and somewhat ambiguous. The factors that most influence decisions are often not concerned with issues of conceptual development and mathematics learning but rather with the exigencies of day-to-day survival. The issues are complex, and their resolution will undoubtedly require more open communication between the groups involved and a reformulation of the major goals of mathematics education.

MANIPULATIVE MATERIALS AND THE ROLE OF THE TEACHER

A recent survey (Weiss 1978) suggests ". . . very common use of an instructional style in which teacher explanation and questioning is followed by student seatwork on paper and pencil assignments . . ." "The NSF case studies (Stake and Easley 1978) confirm this pedestrian picture of day-to-day activity in mathematics classes at all grade levels." (Fey 1979, p. 12)

Systematic use of manipulative materials can have profound effects on the role the teacher assumes in the teaching-learning process. Perhaps most important, teachers must modify their image of being considered the source from which all knowledge emanates. The teacher involved with the active learning of mathematics is no longer primarily concerned with teaching as it has been traditionally defined, that is, meaning lecturing, demonstrating, and other forms of explicit exposition. Instead, the teacher focuses attention on arranging or facilitating appropriate interactions between student(s) and materials. This is not to say that all instances of telling behavior are abolished, but rather they tend to be significantly limited. This redefined role can be traced directly to the nature of learning as previously discussed. Since children learn best through enactive encounters, appropriate experiences with materials are relied upon to assist children with conceptual development. This does not obviate the classroom teacher; it will always be necessary not only to arrange the conditions of learning but also to discuss, debrief, and encourage future explorations by asking the right questions or giving an appropriate direction at the most opportune time.

The teacher's role in using manipulatives in a laboratory setting is more complex and in some sense more demanding than the more traditional role of telling and explaining. Individual and small-group work will assume a higher instructional priority. This is usually accomplished with a concurrent deemphasis on the lecture/demonstration. Such a format (a) allows the teacher to differentiate student assignments more realistically (a station approach seems ideally suited for this); (b) frees the teacher to interact with individuals and small groups more extensively, addressing questions and concerns as they arise; and (c) requires a new source of direction insofar as the structuring of student activity is concerned. This is required since the teacher cannot provide direction to all groups simultaneously. Task or assignment cards can fulfill a major portion of this need. These cards are used to define student tasks explicitly. The teacher need only select those that are most appropriate for individuals and/or groups. Since there are literally tens of thousands of these individual assignment cards available commercially, the teacher need not feel solely responsible for creating the activities that children are to undertake.

THOUGHTS ON THE DIMENSIONS OF AN EXPANDED PROGRAM IN MATHEMATICS

Most commercial textbook series are concerned with essentially the same mathematical topics, and these topics are important and should be maintained in the school program. The mode in which these ideas are presented, however, is essentially inconsistent with the psychological makeup of the students. Bruner's three modes of representational thought are basically analogous to the proposition that "children learn by moving from the concrete to abstract." A textbook can never provide enactive experiences. By its very nature it is exclusively iconic and symbolic. That is, it contains pictures of things (physical objects and situational problems or tasks), and it contains the symbols to be associated with those things. It does not contain the things themselves.

Mathematics programs that are dominated by textbooks are inadvertently creating a mismatch between the nature of the learners' needs and the mode in which content is to be assimilated.

The available evidence suggests that children's concepts basically evolve from direct interaction with the environment. This is equivalent to saying that children need a large variety of enactive experiences. Yet textbooks, because of their very nature, cannot provide these experiences. Hence, a mathematics program that does not make use of the environment to develop mathematical concepts eliminates the first and the most crucial of the three levels, or modes, of representational thought.

Clearly an enactive void is created unless textbook activities are supplemented with real-world experiences. Mathematics interacts with the real world to the extent that attempts are made to reduce or eliminate the enactive void. An argument for a mathematics program that is more compatible with the nature of the learner is therefore an argument for a greater degree of involvement with manipulative materials and exploitation of appropriate mathematical applications.

It does not follow that paper-and-pencil activities should be eliminated from the school curricula. However, such activities alone can never constitute a necessary and sufficient condition for effective learning. Activities approached solely at the iconic and symbolic levels need to be restricted considerably, and more appropriate modes of instruction should be considered. This approach will naturally result in greater attention to mathematical applications and environmental embodiments of mathematical concepts.

One way in which this could be accomplished would be to consciously partition the in-school time allocated to mathematics so as to include such things as mathematical experimentation, applications, various logic-oriented activities, and other departures from the status quo. It is unfortunate that a recent study sponsored by the National Science Foundation had to conclude that "elementary school mathematics was primarily devoted to helping children learn to compute" (Stake and Easley 1978, vol. 2, p. 3). This is in contrast to the recommendations of leading mathematics educators (NIE 1975; Post, Ward, and Willson 1977), supervisors (NCSM 1978), and professional organizations (NCTM 1978–1980). These experts generally agree that the "basic skills" in mathematics encompass much more than the mere ability to compute with fractions, decimals, and whole numbers. The expanded definition proposed here has far-reaching implications for mathematics programs. If it is to be taken seriously, it should be noted that the implementation of the recommendations outlined in this chapter would not only result in the students developing a vastly enlarged view of the discipline itself, but would also result in their greater involvement in the learning process. In this event, manipulative materials could effectively assume the dual role of assisting in the development of computational algorithms as well as that of providing an important intermediary between the statement of a problem and its ultimate solution.

There is still much that we do not know about the nature of the learner, the nature of the learning process, and the interaction between the two. Continued study of the nature of the impact of manipulative materials upon conceptual development is needed. Such study should considerably improve our ability to design effective mathematical experiences for children.

POTENTIAL DIRECTIONS FOR FUTURE RESEARCH

Whatever the appropriate role of iconic experiences, it seems clear that the Brunerian model will prove to be overly simplistic since it does not include reference to such variables as the nature and scope of the human interaction patterns that invariably accompany the educational process. Previous research in all areas continually reaffirms the importance of the teacher variable, a variable that has proved to be extremely difficult to identify and control. The research literature regularly suggests that the teacher effect is responsible for the largest percentage of the identifiable variance. This is true regardless of grade level, mathematical topic, or the level of the students' ability. Comprehensive research in the future must surely attend to this difficult area.

Recent interest in the teaching experiment as an alternative to the more traditional form of educational research, which utilizes classical research designs and their attendant statistical analyses, is a promising innovation in research in mathematics education. The teaching experiment is nonexperimental in nature. It typically utilizes fewer students, sometimes omits the use of a control group, and is designed primarily to maximize interaction between investigator and student. In-depth probing of students' reasoning processes is usually the major research objective. Insights gained by the researcher often result in the formulation of new and more precise hypotheses that can at some later point be subjected to experimental research. Important insights into how students of all ages think mathematically have resulted from increased use of this technique over the past decade. In the future such procedures will undoubtedly shed new light on the more subtle and as yet unanswered questions regarding the nature and role of manipulative materials in the learning of mathematical concepts.

To this point research has been designed primarily to address the larger question, "Does the use of manipulative materials produce superior student achievement?" Results thus far have been encouraging.

Research to date has not investigated the nature of the factors surrounding the use of materials that result in superior learning. When these factors have been isolated and clearly identified, it will become important to explore further the kinds of interactions between individual differences, learning styles, teaching styles, the structural nature of the most useful materials, the relationship between content and materials, and the sequencing and appropriate use of various modes of representation.

The magnitude of this task is enormous and will undoubtedly consume a major portion of the remainder of this century. It is not a task that can be effectively undertaken by isolated individuals, as answers to these questions will require large-scale externally funded cooperative research projects. These projects will undoubtedly identify a series of related questions for subsequent investigation. If such questions are identified and the total research package planned so that each question and answer will supply a piece of a larger mosaic, the results can and will begin to answer questions that at this point are still in the formative stage.

REFERENCES

Bruner, Jerome S. *The Process of Education.* Cambridge: Harvard University Press, 1960.

Bruner, Jerome S. *Toward a Theory of Instruction.* Cambridge: Harvard University Press, 1966.

Dewey, John. *Experience and Education.* New York: Macmillan Co., 1938.

Dienes, Zoltan P. *Building Up Mathematics.* Rev. ed. London: Hutchinson Educational, 1969.

Dienes, Zoltan P. "An Example of the Passage from the Concrete to the Manipulation of Formal Systems." *Educational Studies in Mathematics* 3 (June 1971): 337–52.

Dienes, Zoltan P., and Golding, Edward W. *Approach to Modern Mathematics.* New York: Herder and Herder, 1971.

Fennema, Elizabeth. "Models and Mathematics." *Arithmetic Teacher* 19 (December 1972): 635–40.

Fey, James T. "Mathematics Teaching Today: Perspectives from Three National Surveys." *Arithmetic Teacher* 27 (October 1979): 10–14.

Fitzgerald, William M. *About Mathematics Laboratories.* Columbus, Ohio: ERIC Information Analysis Center for Science, Mathematics, and Environmental Education, 1972. ERIC: ED 056 895.

Ginsberg, Herbert, and Opper, Sylvia. *Piaget's Theory of Intellectual Development.* Englewood Cliffs, N.J.: Prentice-Hall, 1969.

Jackson, Robert. "Hands-on Math: Misconceptions and Abuses." *Learning* 7 (January 1979): 76–78.

Jourdain, Phillip E. "The Nature of Mathematics." In *The World of Mathematics,* edited by James R. Newman, vol. 1. New York: Simon and Schuster, 1956, p. 71.

Kieren, Thomas E. "Activity Learning," *Review of Educational Research* 39 (October 1969): 509–22.

Kieren, Thomas E. "Manipulative Activity in Mathematics Learning." *Journal for Research in Mathematics Education* 2 (May 1971): 228–34.

Lesh, Richard A. "Applied Problem Solving in Early Mathematics Learning." Unpublished working paper, Northwestern University, 1979.

Man: A Course of Study. Cambridge: Education Development Center, 1969.

National Council of Supervisors of Mathematics. "Position Paper on Basic Skills." *Mathematics Teacher* 71 (February 1978): 147–52.

National Council of Teachers of Mathematics. *Mathematics Teacher* 71 (February 1978): 147. (Endorsement of NCSM's position on basic skills.)

National Council of Teachers of Mathematics. *An Agenda for Action: Recommendations for School Mathematics of the 1980s.* Reston, Virginia: NCTM, 1980.

National Institute of Education. *The NIE Conference on Basic Mathematical Skills and Learning*, vol. 1: *Contributed Position Papers*; vol. 2: *Reports from Working Groups*. Washington, D.C.: National Institute of Education, 1975.

Piaget, Jean. *The Psychology of Intelligence.* Boston: Routledge and Kegan, 1971.

Post, Thomas R.; Ward, William H., Jr.; and Willson, Victor L. "Teachers', Principals', and University Faculties' Views of Mathematics Learning and Instruction as Measured by a Mathematics Inventory." *Journal for Research in Mathematics Education* 8 (November 1977): 332–44.

Reys, Robert E., and Post, Thomas R. *The Mathematics Laboratory: Theory to Practice.* Boston: Prindle, Weber, and Schmidt, 1973.

Stake, Robert E., and Easley, Jack A., Jr. *Case Studies in Science Education*, vols. 0–15. Champaign-Urbana, Ill.: Center for Instructional Research and Curriculum Evaluation, University of Illinois, 1978). ERIC: ED 156 498–ED 156 512.

Suydam, Marilyn N., and Higgins, Jon L. *Review and Synthesis of Studies of Activity-Based Approaches to Mathematics Teaching.* Final Report, NIE Contract No. 400-75-0063, September 1976. (Also available from ERIC Information Analysis Center for Science, Mathematics and Environmental Education, Columbus, Ohio.)

Vance, James H., and Kieren, Thomas E. "Laboratory Settings in Mathematics: What Does Research Say to the Teacher?" *Arithmetic Teacher* 18 (December 1971): 585–89.

Weiss, Iris. *Report of the 1977 National Survey of Science, Mathematics, and Social Studies Education.* Durham, N.C.: Research Triangle Institute, 1978. ERIC: ED 152 565.

Wilkinson, Jack D. "A Review of Research Regarding Mathematics Laboratories." In *Mathematics Laboratories: Implementation, Research, and Evaluation*, edited by William M. Fitzgerald and Jon L. Higgins. Columbus, Ohio: ERIC Information Analysis Center for Science, Mathematics, and Environmental Education, 1974. ERIC: ED 102 021.

9. Computers and Calculators in the Mathematics Classroom

Ross Taylor

With the rapid evolution of electronic computing technology, considering the implications of this technology for instruction in mathematics is timely. In this chapter, I will first clarify some misconceptions in the commonly accepted myths about the use of computers and calculators in the classroom. Then I will look at trends in technology and curriculum, examine the factors that inhibit the use of calculators and computers in the classroom, and see what needs to be done to take advantage of the potential of computer-assisted instruction (CAI) and what revisions of the mathematics curriculum need to be made due to the evolution of electronic computing. Finally, I shall make some suggestions about what individual schools should do to integrate calculators and computers into their programs.

In this chapter, the term *computer-assisted instruction* is used to include all direct uses of the computer by students, teachers, teacher aides, and parent volunteers. Some of these uses are briefly described below. The number and variety of instructional computer applications is growing rapidly, so that this list is by no means comprehensive.

I want to express appreciation to Sally Sloan, Minneapolis Public Schools; David C. Johnson, University of London; Carole Bagley, Minnesota Educational Computing Consortium; and Tom Hanson, Minnesota Educational Computing Consortium, for their help in providing background information for this chapter.

Drill and practice: The student interacts with the computer in short drills or in a more comprehensive program involving the computer in a management function.

Tutorial: The computer provides direct instruction, and the student responds to questions posed by the computer.

Learning games: Concepts are learned through lessons in game format.

Simulation: From a variety of disciplines, experiences that are too broad, complex, inconvenient, or dangerous for actual exposure to students are portrayed on the computer.

Problem solving: The computer is used to help solve problems in mathematics or other fields.

Programming: Students learn one or more computer programming languages.

Demonstration: The computer is used to demonstrate concepts in mathematics or other fields.

Computer literacy: Students learn what computers can do and what they cannot do.

Computer science: Students learn about the design and functioning of computers.

Computer generation of materials: The computer generates various types of materials, including tests and work sheets along with answers.

Test scoring: The computer scores tests and analyzes the data.

Management of instruction: The computer handles data and provides the teacher with information that is helpful in the management of instruction.

Information retrieval: The computer provides needed information for career planning or other purposes.

CAI is provided by means of stand-alone microcomputers or terminals that are connected to a central time-sharing computer. Time-sharing computers vary in size from those that handle just a few users to those that can be utilized by hundreds of users simultaneously. Most frequently, in a time-sharing system, the computer communicates with terminals by telephone. A standard telephone receiver is inserted into an acoustic coupler that is connected to the terminal. In some installations, when the terminals are in the same building as the computer, they are "hardwired" directly to the computer.

Some microcomputers can be used either as stand-alone computers or as terminals that access time-sharing systems. In situations where the same program is available on a stand-alone microcomputer or on a time-sharing system, the difference is generally transparent to the user.

MYTHS

Myth: Use of Calculators Will Harm Students' Achievement in Mathematics

During the last half of the 1970s, when inexpensive hand-held calculators suddenly became available, parents, teachers, and administrators were apprehensive about using them in the classroom. The greatest concern was that the use of calculators might be detrimental to the students' achievement in mathematics. In particular, there was a fear that students would become so dependent on calculators that they would lose their ability to compute.

The public concern about the use of calculators in schools motivated a number of research studies on the topic. The National Institute of Education funded the Calculator Information Center to serve as a clearinghouse for information about the use of calculators in the schools. The Center is located at Ohio State University and directed by Marilyn Suydam. In May of 1979 she summarized the research up to that time on the effects of the calculator and she reaffirmed her original findings again in 1980.

Almost 100 studies on the effects of calculator use have been conducted during the past four or five years. This is more investigations than on almost any other topic or tool or technique for mathematics instruction during this century, and calls attention to the intensive interest about this potentially valuable tool. Many of these studies had one goal: to ascertain whether or not the use of calculators would harm students' mathematical achievement. The answer continues to be "No." The calculator does not appear to affect achievement adversely. In all but a few instances, achievement scores are *as high or higher* when calculators are used for mathematics instruction (but not on tests) than when they are not used for instruction. Thus, many researchers working with students at all levels have paralleled the conclusion of Shumway and his coinvestigators that "There were no measurable detrimental effects associated with use of calculators for teaching mathematics. (Suydam 1979, p. 3)

Myth: Computers Will Replace Teachers

Over the years, whenever new technology for education has been developed, we have heard from overenthusiastic proponents that the "good" new technology would solve our educational problems by replacing "bad" teachers. We have heard this about motion pictures, educational television, programmed instruction, teaching machines, and we continue to hear it about computer-assisted instruction. However, technology has never replaced teachers. Furthermore, there is no reason to believe that teachers will ever lose jobs to computers.

Anyone who believes that computers can replace teachers does not comprehend the complexity of the teacher's role. Technology tends to replace people in low-level positions that do not require human interaction and decisions. It tends to provide valuable services for persons in professional positions and eliminates the drudge work, thus enabling them to do their work more effectively.

Experience and common sense tell us that computers will not be effective as a replacement for "bad" teachers. However, computers have tremendous potential for helping to make "good" teachers better. The most promising approach would be to analyze what teachers do and then see what computers can do to help them do it better. We should also seek to extend what teachers are doing into new areas that computer technology has made accessible. The teacher's role can change as computers are able to relieve them of some of their low-level functions, such as record keeping or making up work sheets. Rather than eliminating jobs, the introduction of computers tends to create jobs and allows teachers to function at a higher professional level.

Attempts to use computers to replace teachers will fail because the concept is unrealistic and because such attempts will be strongly resisted. Attempts to use computers to help teachers to do a better job are likely to succeed because the concept has great potential and because such attempts will be broadly welcomed.

Myth: To Use Computers, Teachers Must Become Computer Experts

Teachers who do not use computers may conclude that a high degree of expertise is necessary before one can use a computer. This attitude is most likely to arise in schools where instructional use of the computer is limited to teaching programming and computer science or to using the computer as a problem-solving tool in mathematics.

A certain amount of mystique seems to have developed about computers. Unfortunately, many adults have tended to build up a level of "computer anxiety" that must be overcome before they are willing to try using computers. However, computer anxiety does not seem to be a problem with most students, particularly young children. They do not seem to be afraid that they will make a mistake in front of others or that they will do anything that could harm the computer. On the other hand, adults have been conditioned (properly so) to beware of technology whose use they do not fully comprehend. In addition, many adults have built up additional anxiety about anything that appears to be mathematical.

Adults need to learn that they do not have to be able to program a computer in order to use one. Teachers should be made aware that most of the instructional uses of computers in the classroom require no knowledge of programming whatsoever. Adults have no problem performing other general functions involving technology, such as operating tape recorders or making long-distance phone calls, and with a minimum amount of instruction (less than one hour) followed by periodic opportunity to practice, adults can gain confidence and proficiency in using computers.

Myth: Computers Are Only for the Technically Talented

In many high schools, the use of the computer in instruction is limited to one or more courses in computer science or computer programming at the eleventh-grade or twelfth-grade level, and a strong mathematics background is often a prerequisite. In such an environment, people can easily fail to see that the computer offers many possibilities for a wide range of uses by students of all ability levels in all disciplines at both the elementary and secondary levels. This narrow view is reinforced if it is shared by teachers of computer courses and if they tend to be very possessive about the computer resources available to the school.

Myth: Computers Are Only for Remediation

In some schools, particularly at the elementary level, computers are used exclusively in drill and practice or tutorial programs for students who need remedial work in mathematics or communication skills. In such an environment, a stereotyped notion linking instructional use of computers to remediation can evolve. This misconception is avoided if computer installations are arranged so that the facilities can be used by a variety of students in a variety of ways. For example, computer installations that are primarily for remedial drill and practice should also be used for teaching students to program, to solve problems, to run simulations, and to play computerized learning games.

Myth: Computers Are Useful Only in Mathematics

The computer is useful in every discipline. For example, in English there are programs for teaching reading, spelling, and grammar, and for testing reading levels of materials. There is a range of simulations, drills, and learning games for social studies. Some exciting new programs for teaching music take advantage of the capability of computers to generate sound. With new capabilities in graphics, computer-generated art is rapidly expanding. Programs in home economics deal with diets, recipes, and clothes patterns. For physical education and athletics, there are a number of programs to simulate games like golf or football, to monitor the president's fitness program, to handle football and basketball scouting, and to keep records of interscholastic competition. There are many financial simulations and programs for business education. In addition to entire language courses, there are a number of short drill programs for various modern languages. There are programs for industrial arts, and a number of programs simulate experiments in biology, chemistry, and physics. Counselors have found career information and college selection programs to be very valuable.

TRENDS

Faster and Quieter

During the 1960s and 1970s the terminal most widely used was the standard teletype, which clattered out "hard copy" on rolls of teletype paper at the rate of ten characters per second. This teletype is the Model T of the computer terminal industry. It was noisy and slow, but it was reliable and could stand up to the hard use it received in the education environment. For many years it was the least expensive terminal on the market. But by the end of the 1970s, hard-copy terminals emerged that were relatively silent, could print at thirty characters per second, and were price competitive with the ten-character-per-second teletypes. At the same time, less expensive terminals that could be used to produce a display on a television monitor came on the market. The introduction of microcomputers by the end of the 1970s continued the trend toward faster and quieter output.

Think Small

In the past few years, the computer field has undergone a complete turnabout from what could be called the "dinosaur mentality" to what could be called the "flea mentality." During the 1960s and early 1970s the vogue was to put as many applications as possible on one large computer that would supposedly do everything for everybody in the most cost-effective manner. For example, there were numerous attempts to provide administrative data processing services and instructional services using the same computer. This concept has never really worked effectively. Computers tended to perform one or the other but not both of these services efficiently. When compromises had to be made, administrative uses such as getting out the payroll took precedence, and instruction tended to suffer.

During the 1970s there was a trend toward providing administrative services and instructional services from separate computers, with instruction being served by a range of time-sharing computers that could handle between a few and several hundred simultaneous users. By the end of the 1970s, the small stand-alone portable microcomputers entered the market. These microcomputers were suitable for small business use, home use, and many instructional applications in the schools.

The development of microcomputers offers the schools an excellent opportunity to expand greatly the use of computers in the classroom. Whereas terminals used for time-sharing need telephone access, microcomputers only need an electrical outlet to operate. This allows for much greater flexibility of use. Furthermore, the elimination of the telephone expense is a significant saving, particularly in view of the fact that telephone rates continue to rise. With microcomputers, technical problems can have less impact. For example, if a time-sharing computer system "goes down," then all of the terminals that depend on the system become inoperative. However, failure of one microcomputer does not affect others.

Microcomputer users are not affected by other users on the system. For example, a time-sharing system user may call up the computer and receive a busy signal because the system is completely in use by other users. However, use of a microcomputer cannot be preempted by a user from a remote location. Furthermore, when use of a time-sharing system becomes heavy, the system may slow down. On the other hand, the speed of a single terminal microcomputer is affected only by the program of its single user.

In the past, the lack of dependability of computer access (for whatever reason) has tended to discourage many teachers from using computers. For example, a teacher might want to use the computer in a demonstration to introduce a concept in a mathematics class, but then the teacher must be able to depend on the availability of the computer at the time of the class. There is a real problem if the computer is "down" or is filled up with other users at that time. After several unfortunate experiences of this type, a teacher is likely to give up on computers. A teacher with access to several microcomputers will not have to worry about this problem because if one of the "micros" develops problems, then another can be used.

Teachers have much greater control with microcomputers. They can decide what type of micro to buy to meet their needs. Larger computers are generally used to serve a diversity of needs, and compromises have to be made in the selection process. The move to microcomputers for the instructional use is part of an overall trend toward smaller computers to meet the needs of specific users.

Blend Large and Small

Will microcomputers completely replace time-sharing for instructional use? In view of the rapid developments in computer technology over the past thirty-five years, it would be unwise to make any long-range predictions. However, time-sharing will still be needed in the near future. Some programs available on larger systems need more storage, language, or processor capability than is currently available on microcomputers. For example, large instructional management packages or guidance information packages that use a large data base need more capability than is presently available on microcomputers.

Microcomputers will also not replace time-sharing in situations where information must be continuously updated from a number of different locations. The use of computers for making airline reservations is an example of this. This type of use in education is relatively limited at present. As educators become more computer literate, however, such applications may grow. For example, computers can be used for ordering films from the film library of a school system. In the future, the need to conserve energy and to conserve paper could stimulate more of this type of time-sharing application.

Some microcomputers may be used either as stand-alone computers or as terminals for accessing other computers. Creative use of these capabilities can combine the best features of time-sharing and microcomputers in a cost-effective way. For example, the time-sharing system can maintain a library of programs that will run on microcomputers. Programs developed on a micro can be "uploaded" onto the time-sharing system and then "downloaded" onto other microcomputers. This procedure offers maximum availability of programs to micro users, while minimizing communication and time-sharing costs. The procedure offers the greatest economy if the time-sharing computer is accessed during off hours when phone rates and computer rates are lowest.

Go to Graphics

In the late 1970s, at the same time that microcomputers became widely available, graphics capability—including color graphics—also became available on computers at a realistic cost. This graphics capability offers excellent possibilities for creative programming by students, for exciting interactive computer-assisted packages, and for use by teachers in demonstration. For example, concepts involving functions can be presented more effectively and learned more quickly using a graphics display viewed by the entire class on television monitors. In the study of trigonometric functions, the difference between $\sin(x+2)$ and $(\sin x)+2$ can easily be shown by producing their graphs along with the graph of $\sin x$ in different colors on the screen. The teacher can ask the students to predict what will happen to a function if a certain parameter is changed. Then through projecting the function on the screen, the prediction can quickly be checked without the need for laboriously plotting the function point by point on the chalkboard or on an overhead. In addition to color graphics, the sound generation capabilities of newer computers open up a number of new possibilities for CAI.

Forward to the Basics

The computerization (and "calculatorization") of our society is producing two types of changes in the mathematics curriculum: (1) changes to prepare students better for living in our computerized society and (2) changes to improve instruction that are possible because of the availability of computers and calculators for classroom use.

Some of the changes needed to prepare students for the world in which they live as adults are set forth in the National Council of Supervisors of Mathematics (NCSM 1978) "Position Paper on Basic Mathematical Skills." The paper lists ten important basic skill areas, one of which is computer literacy. (See Chapter 13 of this book for a more complete discussion of this position paper.)

Instead of going back to the basics, the NCSM position makes the point that in mathematics we must go forward to the basics essential to adults in the present and future. Primarily because of computers and calculators, the skills needed for the present and the future are different from those needed in the past.

We know that we face a world of change in which we will continually encounter many different kinds of problems that are new to us. For that reason, the role of problem solving is coming to the forefront in mathematics as well as in other disciplines. For example, the NCSM position paper states that "learning to solve problems is the principal reason for studying mathematics" (NCSM 1978, p. 148). Availability of computers and calculators is contributing to the increasing importance of problem solving, and their usefulness as problem-solving tools is causing a restructuring of priorities in mathematics education. Problems that were, for all practical purposes, impossible to solve are now possible through the use of electronic computing. For example, calculating the number of seconds in a year becomes easy for an upper elementary student with a calculator, and producing a list of prime numbers between 1 and 10,000 becomes realistic for a secondary student with access to a computer. Furthermore, access to computers and calculators opens up a rich new source of problems at the school level. For example, it is now realistic for students who have not studied logarithms to ask how many years it will take until a twenty-cent candy bar costs $1,000, assuming a given rate of inflation.

Today, computation must be viewed from a new perspective. The ability to do rapid mental calculations is an important skill to have in setting up a problem and exploring possible solutions. Mental calculation is also valuable for checking the reasonableness of results. Today, long and complicated calculations are accomplished using a calculator or a computer. However, calculators are too slow, too cumbersome, and not always accessible for simple calculations. For example, who wants to have to find a calculator, turn it on, and then press four different buttons just to obtain a particular multiplication fact? Students still need to learn to do paper-and-pencil calculations, both as an alternate method to the calculator and to handle situations for which calculators are not suitable. For example, most calculators can not handle computation involving fractions.

Calculators and computers have created a need for a dramatic shift in emphasis in computation. More stress should be placed on decimals and less on fractions. (Conversion to the metric system is another reason for emphasizing decimals.) Instruction in computation with decimals can, and probably should, precede computation with fractions, provided that basic concepts of the meaning of fractions have been taught. Students should still learn how to compute with fractions, but complicated problems involving hard-to-find common denominators should be avoided. Ability to handle easily a variety of simple computations with whole numbers, decimals, percents, and fractions should be stressed. Repetitive work involving cumbersome computation such as long division with multidigit divisors should be avoided.

In a world that relies more and more on electronic computing, estimation and approximation skills take on increased importance. Students should learn to combine approximation skills and simple computation skills to arrive at estimated solutions to problems. Then they will be able to determine if their electronically computed results are reasonable.

Toward Computer Literacy

Computer literacy is listed as one of the ten vital basic mathematical skill areas in the NCSM position paper. The phrase *computer literacy* has a wide range of interpretations from simple awareness of the power and the limitations of computers to comprehensive knowledge of computer hardware, history, applications, and operation, as well as ability to program in one or more computer languages. As a minimum, all citizens should have knowledge of the capabilities and limitations of computers, and they should realize how the use of computers affects their lives. They should learn to challenge statements like "The computer made a mistake" or "The computer won't allow us to do it that way."

With the proliferation of computers, more and more people will want to learn how to operate them for their own personal use or for their jobs. Knowledge of computer programming and other aspects of computers will become increasingly important for personal use and for career opportunities.

Some may question why computer literacy is considered a basic mathematical skill area. They might suggest that computer awareness fits more appropriately within the social studies curriculum. From a logical point of view, social studies would be the appropriate place to deal with the social impact of the computer. However, the computer has been largely avoided or ignored by the social studies discipline. Currently, the mathematics discipline offers the best opportunity because—in most cases where computers are used in schools—the leadership has come from the mathematics department. As a practical matter, if the mathematics education profession does not take the initiative in pushing for instruction in computer literacy, then it probably will not get taught.

The idea of a semester course in computer literacy has been proposed. Such a course might be possible as an elective; however, any attempts to require such a course are not realistic in view of our crowded curriculum. A more promising approach is to include computer literacy units in mathematics as well as in other courses. Widespread use of computers in instruction will facilitate computer literacy.

INHIBITING FACTORS

Factors Inhibiting the Use of Calculators

No one questions the rapid emergence of inexpensive hand-held calculators, but the use is not as widespread as one might expect. This is partially due to the following factors.

1. *Appropriate curriculum:* The use of calculators has produced a need for curriculum revision. The Conference on Needed Research and Development on Hand-Held Calculators in School Mathematics, sponsored in 1976 by the National Institute of Education (NIE) and the National Science Foundation (NSF), produced seven recommendations for curriculum development (NIE and NSF 1976, pp. 18–19).
2. *Appropriate teacher education:* Recommendation 18 of the NIE-NSF Calculator Conference states:

   > Training and retraining of teachers to help them respond to calculators and calculator-influenced curriculum materials must be made equal in importance to the development of initiatives aimed at school-age children. (NIE and NSF 1976, p. 20)

   Teachers are more ready to use calculators than computers because they already use calculators at home, and calculators do not have the "mystique" of computers.
3. *Cost:* In spite of tight school budgets, calculator costs have declined to the point that they should be available to all students. In some situations, schools can ask students to supply their own calculators, with the schools supplying them only as needed. In other cases, where the schools must supply them, a little imagination can locate funds. For example, calculators can be purchased

with Title IV-B funds that are available to all public schools, and PTAs and other groups that make donations to schools actually prefer to use money for tangible items like calculators.

4. *Logistics:* When hand-held calculators first became widely available, there was great concern that they might be stolen. However, experience has indicated that in most cases security is not a major problem.

   The problem of the energy supply for school calculators has been reduced with the development of battery-operated calculators that function for hundreds or thousands of hours on the one battery. Reliability of calculators has increased significantly over the past few years, and maintenance is not a factor because at today's prices calculators are less expensive to replace than to repair.
5. *Fear that calculators will be harmful to learning:* As indicated in the review by Suydam, there is a growing body of evidence that this fear is unfounded.
6. *Resistance to change:* Teachers are in the key position to effect or resist change at the school level. If teachers are provided with appropriate training and support, they will be likely to implement the use of calculators in their classrooms.

Factors Inhibiting the Use of Computers

In 1971 the National Science Foundation funded a study by EDUCOM to identify hindrances to the use of computers in instruction (Anastasio and Morgan 1972). I was one of twenty "authorities" who filled out the questionnaire and participated in the final conference of the study. Eight years after this study, computers are still not being used in education as widely as many had expected. The study identified six inhibiting factors; these are listed below along with comments on their effects at the beginning of the 1980s.

1. *Production/distribution of instructional materials:* The participants in the EDUCOM conference identified this as the most critical area. I did not agree, feeling that the cost factor was the most critical. I believe that the reason for our difference was that while most of the experts were producers of computer materials, my role was primarily that of a consumer.

   As publishers and computer companies perceive that there is a market in computer materials for instruction, the materials will be produced and distributed. Funding from the federal government, states, and private foundations, however, is often needed for the initial development of particular types of materials before a commercial market can be established.

2. *Demonstration:* Participants in the conference pointed to the lack of compelling evidence that computer-assisted instruction is more effective than other instructional methods. In the absence of such evidence, how can the costs of computer-assisted instruction be justified? Recent reviews of research have addressed this question.

   A review by Taylor et al. (1974) of studies on the effectiveness of CAI indicated that CAI tutorial, drill, and practice activities were effective at the elementary level, particularly for low-achieving students. A review of studies on the effectivenes of computer-assisted instruction in secondary schools (Thomas 1979) concluded that CAI leads to achievement levels equal to or higher than traditional instruction, as well as to favorable attitudes, significant time savings, and comparable levels of retention and cost. Johnson (1978) cited several studies indicating that studying mathematics with computers improves problem-solving capability.

3. *Theory of instruction:* Conference participants noted that many curriculum developers failed to recognize that materials must be restructured for use with computer systems. For example, they decried use of the computer merely as an electronic page turner for programmed instruction. (The only visible advantage to such a procedure is that it eliminates page-turning fatigue.) Participants also noted the lack of an adequate range of computer-based pedagogical techniques, such as question answering, simulation, gaming, and problem-solving systems.

   I was not as concerned about this point as the others. At the turn of the decade, we still do not have sufficient theory to satisfy the theoreticians, but we practitioners must continue to provide instruction every day, sufficient theory or not. We can still make progress if we recognize the complexity of the teaching-learning process, view the computer as a support to rather than a replacement for the teacher, and continue to explore a variety of human-computer interactions in school settings.

4. *Educational system and the teacher:* A significant obstacle was the reluctance of school personnel to carry out the reorganization and training that broad use of CAI would entail. Implementation of

CAI encounters the same difficulties as implementation of other innovations. Innovations are often conceived by experts outside the schools. Then an attempt is made during implementation to restructure the school to fit the innovation. The result is often failure, and the experts cannot understand why the school personnel do not embrace the innovations. An approach that starts in the schools and involves school personnel from the beginning is usually much more successful at the implementation stage. Reluctance to change is a significant deterrent, but it can be overcome by structuring the situation so that school personnel have a vested interest in the change.

The provision of in-service and preservice training of teachers is essential. School systems and institutions of higher education need to increase their efforts in this area. Quality in-service training can also be provided through computer consortiums or intermediate school districts that serve a number of local school districts.

5. *Cost:* In the past, cost has been the most significant factor inhibiting the use of CAI. However, while the cost of everything else has been going up, the cost of computing has been coming down. The advent of the microcomputer provides a significant cost breakthrough that should make the greatly expanded use of computers in instruction possible. A large initial capital investment is no longer necessary to initiate computer-assisted instruction.
6. *Technical research and development:* Technical problems were not judged to be as important a deterrent as the other problems. Most computer hardware used by the schools is also used in business, industry, and—more recently—in the home. Education should continue to benefit from increasingly rapid technical developments.

WHAT NEEDS TO BE DONE?

We Need to Use the Potential of CAI

With improving computer technology and decreasing hardware costs, the potential for improving instruction with the help of the computer is rapidly expanding. Therefore, the National Science Foundation, the National Institute of Education, and the United

States Department of Education should make research and development in the area of computer-assisted instruction a priority. The more school based these projects are, the more likelihood they will have for success. The payoff will be greater if efforts are directed toward helping teachers to improve instruction rather than trying to replace teachers. This approach will generate teacher support, which is critical to successful implementation.

The use of a computer and a video screen with graphics to demonstrate mathematical concepts to a class offers exciting possibilities. The capability of the computer to generate tests, work sheets, and other learning materials should be fully exploited. A computer with a card reader or another type of optical scanning input device can score tests, analyze the results, and keep records. Teachers who use a systematic approach and teach to objectives could effectively use the computer to help them manage instruction. Computers can be used interactively for giving tests or for providing instruction in areas where the computer can be particularly effective. The objective should be to blend the computer successfully into the teaching-learning process. Drill and practice should be provided as needed for skill maintenance. The priority should be for using CAI to help students learn concepts with which they have the most difficulty. The use of CAI to make a wider range of courses available to students in small schools in remote locations should also be pursued.

Appropriate teacher training is vital to effective CAI implementation, so the National Science Foundation and other agencies need to support such efforts. Higher education, state departments of education, and local school systems should all take responsibility for seeing that teachers have opportunities for in-service training.

State governments can also play a significant role in stimulating growth in computer-assisted instruction. For example, the Minnesota Council on Quality Education has supported several computer curriculum development projects. The Minnesota Educational Computing Consortium provides instructional time-sharing service to all schools in the state. It also provides other computer support services, including curriculum development and in-service and resource support through regional coordinators. There are a number of examples throughout the country where computer services are provided to a number of schools through a computer consortium or a county or regional service unit. These types of service organizations have demonstrated great potential for improving and expanding the use of CAI.

However, schools participating in these organizations must be careful to maintain control. Service organizations must function as servants, not masters.

We Need to Revise the Mathematics Curriculum

The rapid proliferation of electronic computing power is one of a number of factors that makes a total revision of the mathematics curriculum desirable or even necessary at this time. Other factors include conversion to the metric system, pressures for increased student competency, our increasing knowledge of how children learn mathematics, an increasing awareness of the debilitating effects of sexism and racism in mathematics, and increasing mathematics requirements for entering an ever-widening range of careers and higher education programs. Currently, our society has a shortage of persons with mathematical expertise that can be applied in vastly diverse careers. We must stimulate students to take all the mathematics they can handle. To accomplish this, we need informed counseling, inspired teaching, and an improved curriculum. The curriculum improvements must focus on providing students with basic skills needed by all citizens in modern society and with additional mathematical skills that students need to keep their options open for careers and for opportunities in higher education.

There is a striking contrast between the influence of calculators and that of the metric system on the curriculum. All those involved with mathematics education, including publishers, have predicted the long-awaited conversion to the metric system. Publishers have incorporated the metric system into the standard learning materials, and an abundance of metric activity kits and supplementary metric materials have been produced. Yet conversion to the metric system has not come as rapidly as many had thought, and there are even some who feel that mathematics education was premature in its emphasis on the metric system.

On the other hand, the widespread availability of inexpensive hand-held calculators came so rapidly during the middle 1970s that the mathematics education profession was not prepared. The initial question of "Should children be allowed to use calculators?" was quickly resolved. Today, most children have access to calculators at home. (For example, the 1978 Minnesota State Assessment found that over 90 percent of the students have access to calculators.) Furthermore, research studies have not found calculators to be harmful, and the mathematics education profession has encouraged appropriate use of the calculator in the classroom. The National Council of Teachers of Mathematics publication, *Position Statement on Calculators in the Classroom* (September 1978) begins:

> The National Council of Teachers of Mathematics encourages the use of calculators in the classroom as instructional aids and computational tools. Calculators give mathematics educators new opportunities to help their students learn mathematics and solve contemporary problems. The use of calculators, however, will not replace the necessity for learning computational skills. (NCTM 1978)

Now the question is not if the calculator should be used in the classroom but how it should be used. For example, calculators can be used to check answers to computation problems, but use of an answer sheet is faster. On the other hand, using a calculator for exploration in learning concepts and solving problems offers exciting new possibilities for instruction. Conferees at the 1976 NIE-NSF Calculator Conference pointed out the conflicts that arose because of the swift development of calculator technology.

> Educators are faced with a dilemma. Their experience and instincts tell them to research, test, and proceed with caution. Yet calculator technology is progressing rapidly, and marketing pressures are great. The evolutionary pace traditionally associated with curriculum change is too slow to fit the present situation. (NIE and NSF, 1976, p. 3)

Some curriculum improvements can be accomplished through direct interaction between the schools and the publishers. For example, when schools have refused to purchase learning materials that are sexist or racist, the publishers have made the necessary revisions to insure that women and minorities are suitably represented and portrayed. However, participants in the NIE-NSF Calculator Conference concluded:

> The education community, especially mathematics educators, must accept a leadership role in helping schools adjust to the existence of calculators. It must lead in delineating curriculum applications of hand-held calculators. It must not default, allowing manufacturers and publishers to make most of the crucial decisions. (NIE and NSF 1976, p. 5)

Research and development efforts of national scope are needed with respect to both calculators and computers. These efforts will need federal financial support and possible private funding as well. Issues dealing with the calculator appear to be more urgent because of the virtually universal availability of calculators. However, the impact of calculators and computers on the curriculum is interrelated, and any efforts should take both into consideration. Electronic computing presents greater need and opportunity for curriculum development in areas such as problem solving, consumer applications, estimation and approximation, computer literacy, algorithm design, iterative procedures, and mathematical modeling.

The curriculum changes cannot be mere add-ons to the existing curriculum. The current curriculum must be examined carefully to determine which topics can be deemphasized or deleted and which can be taught more efficiently through use of calculators and computers. Some restructuring and reordering of the curriculum may be in order. The 1976 NCTM position on computers in the classroom states:

> The astounding computational power of the computer has altered priorities in the mathematics curriculum with respect to both content and instructional practices. . . . An essential outcome of contemporary education is computer literacy. . . . Educational decision makers, including classroom teachers, should seek to make computers readily available as an integral part of the educational program. (NCTM 1976)

Concerted efforts at the federal, state, and local levels are needed to integrate electronic computing fully into the mathematics curriculum.

We Need to Revise Mathematics Testing

There is great emphasis today on accountability, competency, and test scores; in this environment, testing has a large influence on the curriculum. In most cases school programs are evaluated on the basis of standardized norm-referenced tests. The choice of tests is limited to the products of a handful of publishers, and none of the standardized tests currently in existence or in the planning stage allows for the use of calculators. A major recommendation of the National Advisory Committee on Mathematical Education (NACOME) in 1975 was:

> that beginning no later than the end of the eighth grade, a calculator should be available for each mathematics student during each mathematics class. Each student should be permitted to use the calculator during all of his or her mathematical work including tests. (NACOME 1975, p. 138)

For this recommendation or a similar recommendation to be implemented, tests that are different from the standardized tests in use today will have to be employed. The test publishers will probably not produce the types of tests needed until they can see that there is a market for them. In the meantime, if progress is to be made, tests that fit the changed objectives of the curriculum will have to be developed. This can be accomplished through local efforts, through nationally funded efforts, or through state-funded efforts. With the emphasis at the state level on assessment and on competency in basic skills, the states could provide leadership in developing tests that are more in line with current needs.

One problem with locally developed tests is that they do not provide information about how local students compare in performance with students from other areas—something the public wants to know. If the national assessment model is used where comparisons are made on an item-by-item basis, then local schools can select items from the national or state assessment that fit local objectives. Then, for example, computer literacy objectives can be tested. Use of calculators can be allowed if they were allowed when the data on students' performance on the test items were gathered.

WHAT SHOULD SCHOOLS DO?

What Should Schools Do about Broadening the Use of CAI?

Everyone in education should realize that computers have application in all disciplines at all levels of instruction for students of all ability levels. Everyone benefits if persons using computers actively strive for their broadest possible use. Parochialism and possessiveness with respect to computers are self-defeating. For example, it may happen that in a school the computer operation may become the sole province of a teacher or a small group of teachers, often in the mathematics department. This can happen to the extent that the school's computer facilities become known as an individual teacher's personal facilities. Then the entire school can get the notion that computers are only for very restricted instructional use, such as the teaching of computer programming. Under these circumstances, the teacher or small group of teachers are unlikely to get much support from the other faculty members for expanding the computer facilities at budget time. However, if an attempt is made to share these facilities and to widen computer use as much as possible, then broad faculty, parent, student, and administrative support for expanding the *school's* computer facilities can be generated. Furthermore, the students will benefit more fully from the wide range of computer applications available in the school.

What Should Schools Do about Electronic Computing in Mathematics?

Educational decision makers should move to integrate calculators and computers fully into the total mathematics curriculum for kindergarten through grade twelve. Special courses could be offered, but the major thrust should be in revising the standard mathematics program.

At this time, computer technology is changing at an ever-increasing rate. Bork suggests how the administrator and the faculty member can react to this rapid change.

> The key idea is flexibility. It will be impossible to make decisions that will last for twenty-five years or even two years. Plans developed each five years will be inadequate. A continual assessment within the institution will be required to determine how the institution can best respond to the situation. The assessment process must clearly involve a range of people including the faculty most interested in exploring the uses of the computer in instruction. (Bork 1978, p. 133)

In light of the rapid change, any of my recommendations concerning types and amounts of hardware that schools should have must be very tentative. Calculators should be available to every student with the possible exception of the earliest elementary grades. At the lower levels, simple four-function calculators are sufficient. Beginning during junior high school, a square root function would be desirable. Calculators used in upper-level high school courses should have the logarithmic and trigonometric functions and their inverses. Preferably, there should be a sufficient supply of calculators so that each student has one to use. Students should be encouraged to use their own calculators whenever feasible. Electronic calculatorlike devices for drill and practice and other mathematics activities should be obtained for suitable uses, and calculator use should be controlled by the teachers. For example, use of a calculator would not be allowed during a test on computation.

Every secondary mathematics class should have access to a computer with a video screen for classroom demonstration. In many cases the teacher will want the computer and screen permanently assigned to the classroom. Each secondary school should have the capability of providing access to computers on a ratio no higher than three students to one stand-alone computer or computer terminal. Additional computers or terminals should be provided as needed. In addition to stand-alone microcomputers, schools should have access to a time-sharing system. In addition to video screen output, schools should have hard copy output capability. Card readers or other optical input capability should also be available. Ample computer facilities should be available in all elementary schools.

Schools should approach CAI with the concept that computers are for use in all subjects at all grade levels by all students and all teachers. With respect to the location and the monitoring of computers and terminals, practices will vary considerably from school to school. Equipment can be assigned to departments or teachers on a permanent basis or on a checkout basis. We can expect to see more computer laboratories and more computer equipment in media centers.

The need for computer support personnel goes along with computer use to give teachers, teacher aides, and parent volunteers the help they need to implement the widest and most effective use of computers for instruction. Ideally, this type of support service should be provided at the school district level, at the computer consortium level, and at the local school level. At the school level, the service will probably be provided on a part-time basis by a knowledgeable faculty member. These support personnel can arrange for in-service opportunities to meet local needs. With the rapid increase of educational computer use, this service will be even more important than the service of audiovisual coordinators has been in the past.

What Should Schools Do about Buying and Supporting Instructional Computer Hardware?

Many other questions need to be raised and answered in order to resolve this question.

1. *What instructional needs will be addressed through use of the equipment?* A survey of faculty should reveal many of these needs. Additional information can be gained from computer support personnel.
2. *What type of equipment should be purchased?* Will the educational needs be met best by use of terminals for access to time-sharing systems, microcomputers, or hand-held electronic devices? One needs to see what programs are available to meet the educational needs and on what type of equipment they are available. Then the most cost-effective alternative can be selected. For example, if the need addressed is one of providing information to students about career and higher education choices,

then one will want to select a terminal to access programs that are only available through time-sharing. If the need is for drill and practice on basic number facts, then perhaps the most cost-effective alternative would be the relatively inexpensive hand-held electronic device. If the instructional need is to teach computer programming, then a microcomputer might be the most cost-effective alternative. Due to the rapid changes in technology, one should check with a computer support person who is keeping current on the state of the art before making a final choice.

3. *What type of input should the equipment accept?* Most of the computer devices in use have keyboard input. Some of them have provisions for input from paper tape or from magnetic tape cassettes or magnetic disks. Some receive input from touch-sensitive screens. If an application like test scoring or attendance monitoring is to be considered, then equipment capable of accepting input via punched or mark-sense cards is necessary.
4. *What type of output should the equipment produce?* If the output is to be via the video screen, should it be black and white or must it be color? For applications like computer generation of materials, hard copy is necessary. Many of the terminals in use today use heat-sensitive paper, which produces nice copy but is quite expensive. Is a ten-character-per-second output (such as that of the teletype terminal) sufficient or must there be capability for output of thirty characters per second or even higher? Must the output be silent or nearly so?
5. *How large a processor is needed?* This question is answered by determining what types of programs are desired and whether or not the computer is capable of handling them. For example, some of the least expensive microcomputers have very small processors that can handle only a limited range of programs.
6. *What computer languages are available?* Virtually all of the computers available for instruction today use BASIC computer language. However, each manufacturer uses BASIC with its own "dialect," and the programs do not always transfer readily from one manufacturer's equipment to another. Usually, some adjustments have to be made in the process. Furthermore, different equipment from the same manufacturer may accept different levels and versions of the language. For example, the least expensive microcomputers accept only a very limited version of the BASIC language. If other computer languages such

as COBOL, FORTRAN, or PASCAL are desired, then a computer that can handle the desired languages should be selected. In microcomputers, a consideration is whether or not the language is built into the machinery or is external to the machinery. If the language is external to the machinery, then there is potential for loading different languages into the computer from external storage devices.

7. *How much and what type of storage is needed?* Many of the microcomputers have provisions for external storage of magnetic tape cassettes or disks. If such storage is used, then a method of storing the cassettes or disks must also be considered. The amount and type of storage will be determined by the size and type of program to be accessed.
8. *Where will the equipment be located?* The equipment must be located so that it is accessible to the faculty and to the students who will use it. Other people should not be distracted by the use. For example, a noisy teletype terminal in a classroom or a media center can be a distraction. Furthermore, the equipment must be located so that its use can be adequately supervised. Sometimes mobility of equipment is desired to meet varying needs for the equipment. In most cases, heavier equipment such as terminals should be mounted on wheels.
9. *How will the equipment be maintained?* Before purchasing the equipment, the duration and the conditions of the warranty should be checked. If one has a choice, one should purchase equipment for which local service is available. If components must be mailed away for service, spare components should be available. Budgetary provisions should be made for maintenance, and maintenance should be monitored to see where a service contract or a parts and labor arrangement is more cost effective.
10. *What type of interface with other equipment is possible?* Some microcomputers can also be used as computer terminals that connect by phone with time-sharing systems. Microcomputers that can "down-load" programs from central computers have an obvious advantage. The possibility for connection and interface with various input, storage, and output devices is a matter to be considered.
11. *If a terminal is to be purchased, how will the computer service and phone service be provided?* Check with the providers of the time-sharing service to be sure that the terminal and acoustic coupler are

compatible with the computer or computers providing the service. Make budgeting provisions for computer service and telephone service.

12. *What is the expected life of the equipment and what is its potential usefulness over that life?* The expected life of equipment is best determined by its "track record." Consequently, for equipment that is brand new on the market, projections are highly speculative. There is always concern about how long the manufacturer will continue to produce and support the equipment. There is also a consideration of potential upgrading of the equipment by adding additional storage, input, or output devices. Inexpensive equipment that does a limited number of things very well can be a good buy because it should continue to do that limited number of things quite well. More expensive equipment should have flexibility and adaptability to meet new needs as developments occur.
13. *How will in-service training be provided?* When new equipment is purchased, there must be opportunities for teachers, aides, parent volunteers, and students to learn how to use it. Provisions for appropriate training should be made.
14. *How will use of the equipment be monitored?* If the equipment is to be used by different teachers, who will establish the priorities and schedule it for teacher use? How will the priorities and schedules for student use be established? Who will be responsible for maintenance of the equipment, including the actual performance of simple maintenance such as the changing of the ribbon on a teletype?

CONCLUSION

There is a wide range of applications for computer-assisted instruction, including drill and practice, tutorial, learning games, simulation, programming, demonstration, computer literacy, computer science, computer generation of materials, test scoring, management of instruction, and information retrieval. The use of calculators has not been found to hurt student achievement in mathematics. In many cases, the use of calculators has helped achievement. Computers will not replace teachers; computers can help increase the effectiveness of teachers. Teachers do not have to be experts to handle most of the instructional programs on computers, and computers are useful for all ability levels and students at virtually every grade level. Computers are useful in every discipline.

There is a trend toward faster and quieter computer terminals, and the use of microcomputers is expanding rapidly. A combination of cost effectiveness and the full potential of CAI is being pursued through a blend of time-sharing and microcomputers. Color graphics and sound generation capabilities of newer computers offer many possibilities for CAI. Calculators and computers are stimulating a forward-looking approach to basic skills in which priorities are changing. Areas like problem solving, approximation and estimation, and computer literacy are increasingly important.

As the price of electronic computing decreases, the cost factor is less and less a hindrance to the use of calculators and computers in the classroom. We will be able to take advantage of the potential of computers and calculators in the classroom if we put priority efforts into research and development and in-service training. It is time for a major revision of the mathematics curriculum and mathematics testing to keep pace with the rapid progress in electronic computing.

Individual schools should implement as broad a use as possible of CAI in all disciplines at all levels. Use of the calculator and the computer should be fully integrated into the total mathematics curriculum, and necessary hardware should be made available. Because of the rapid changes in the CAI field, computer support personnel should be available to the schools. Electronic computing will help teachers to provide better education if the decision makers at all levels of education recognize and support its promise while keeping an eye on cost effectiveness.

REFERENCES

Anastasio, Ernest J. and Morgan, Judith S. *Factors Inhibiting the Use of Computers in Instruction.* Washington, D.C.: EDUCOM, 1972.

Bork, Alfred. "Instructional Systems: Technical Issues." In *Closing the Gap between Technology and Education.* Proceedings of the 1977 EDUCOM Fall Conference (Boulder, Colo.: Westview Press, 1978.

Johnson, David C. "Calculators: Abuses and Uses." *Mathematics Teaching*, no. 85 (December 1978): 50–56.

National Advisory Committee on Mathematical Education. *Overview and Analysis of School Mathematics.* Washington, D.C.: Conference Board of the Mathematical Sciences, 1975.

National Council of Supervisors of Mathematics. "Position Paper on Basic Mathematical Skills." *Mathematics Teacher* 71 (February 1978): 147–52.

National Council of Teachers of Mathematics. *Position Statement on Computers in the Classroom* Reston, Va.: The Council, 1976.

National Council of Teachers of Mathematics. *Position Statement on Calculators in the Classroom.* Reston, Va.: The Council, 1978.

National Institute of Education and National Science Foundation. *Report of the Conference on Needed Research and Development on Hand-Held Calculators in School Mathematics.* Washington, D.C.: National Institute of Education and National Science Foundation, 1976. ED 139 665.

Suydam, Marilyn N. *The Use of Calculators in Pre-College Education: A State-of-the-Art Review.* Columbus, Ohio: Calculator Information Center, Ohio State University, May 1979.

Taylor, Sandra et al. "The Effectiveness of CAI." Paper presented at the Annual Convention of the Association for Educational Data Systems, Arlington, Va., 1974. ERIC, 1974.

Thomas, David B. "The Effectiveness of Computer-Assisted Instruction in Secondary Schools." *Association for Educational Data Systems Journal* 12, no. 3 (1979): 103–16.

PART FOUR
Students and Teachers

10. Attitudes and Mathematics

Laurie Hart Reyes

Mathematics is considered by many people to be an area of study that is objective and removed from feelings. Some people believe that one of the main purposes of studying mathematics is to help develop an orderly, analytical way of thinking. If mathematics is really so far removed from the realm of feelings, why even discuss attitudes or feelings about mathematics? To begin answering this question consider the following teachers' seventh-grade mathematics classes.

Teacher A

Teacher A begins class with a review of some of the previous day's lesson. He commands the attention of his students by moving and speaking quickly. He questions students in a rapid staccato style, accepting as correct only the answers that exactly fit his viewpoint. When students give answers Teacher A considers wrong, he shows annoyance and displeasure. Students caught not paying attention are sometimes chastised humorously but loudly and sometimes chastised only loudly. Teacher A really attempts to stretch students' thinking beyond rote learning with his fast, probing questions. His class is alert and well behaved, but some students seem very intimidated by his style.

Teacher B

Teacher B also begins class with an oral review of the previous day's work. The attention of the class is on the material. He involves most students in the interaction. He asks questions of students but stays with a student until that student understands. Students are given time to think before responding even if the class must wait a few seconds. Teacher B makes an effort to use each student's answer, whether right or wrong, in some productive way. Incorrect answers are not looked upon with disdain. When students do not pay attention, they are not made fun of in front of the class. They are brought firmly back. Teacher B's class is on task, and few students misbehave. Students appear to be confident and willing to bring up questions.

Both of these teachers are effective in their presentation of mathematics. The cognitive environment in both classrooms is one that appears conducive to students' working on mathematics. It is the *affective* environment that varies so much between the two. More students in Teacher A's class seem anxious and hesitant to offer ideas than in Teacher B's class where, indeed, students do seem more willing to learn. Not only do the students appear more comfortable in Teacher B's class, but Teacher B seems to enjoy his class more than Teacher A does.

Although we must be careful not to draw too many conclusions from brief vignettes, it is safe to say that classroom environment is important from the viewpoints of both the teacher and the student. For students, the mathematics class could be a time to dread or a time to look forward to. Certainly we would wish to develop the latter. Students who feel somewhat comfortable in class are undoubtedly easier to teach than students who are blocked by fear, and if students are comfortable, the job of the teacher is more pleasant.

The purpose of this chapter is *not* to argue that the major goal of mathematics education is for all students to like mathematics. Some students, however, are prevented from learning as much as they might by negative attitudes and feelings about mathematics. This chapter argues that an awareness of students' attitudes and their effect on the learning of students is important for teachers. This is particularly valid in light of the growing need for mathematics in our lives.

Knowledge of mathematics is increasingly important for students in order to keep a wide variety of career options open. In fact, mathematics has been called the critical filter for entrance into high-paying, high-prestige careers. There is a strong connection between the number of years of mathematics required for a career and the average annual income for people who pursue that career. Jobs that require a lot of mathematics at entry level tend to lead to higher paying jobs than entry-level jobs that require little or no mathematics.

Unfortunately, in high school many students decide to take only the mathematics courses required for graduation or entrance into college or other post high school training. At the college level many students avoid majors that require more than minimal mathematical competence. By limiting the number of mathematics courses taken in high school, technical school, college, or university, students also limit their possibilities for the future. Many students meet only minimal requirements for entrance into a career and later find that their advancement to higher positions is limited by a lack of mathematical knowledge.

A variety of factors influence students as they make decisions about how much mathematics to take in high school and postsecondary school. Research shows that attitudes or feelings about mathematics are important factors in student decisions. Since their decisions about enrollment in mathematics courses are important ones and student attitudes toward mathematics affect these decisions, an understanding of these attitudes toward mathematics is essential.

Not only do attitudes toward mathematics influence a student's willingness to enroll in more mathematics courses, but these attitudes also influence how much effort a student will put into learning mathematics after enrolling in mathematics classes. Beginning in the elementary school years, continuing into the middle school years, and increasingly in the high school years, students who have positive feelings about mathematics exert more effort, spend more time on tasks, and are more effective learners than students with poor attitudes.

The attitudes of students are also important factors in the classroom environment and in students' motivation. In addition, students' attitudes have been found to affect teachers' treatment of students. Teachers seem to pay more attention to students who are sure of themselves in mathematics than they do to students who are less sure of themselves, even when both sets of students perform equally well in mathematics. So students' attitudes toward mathematics can actually make a difference in how frequently teachers interact with them. This is a particularly important aspect of student attitudes toward mathematics.

Teachers spend considerable amounts of time and energy assessing the cognitive aspects of student performance in mathematics, but they rarely assess student attitudes in any objective manner. Their subjective perceptions of student attitudes often differ considerably from the students' responses to objective measures of attitudes. Teachers believe that certain students feel positive and some negative about mathematics. Yet when an objective measurement is made of these same students' attitudes, it often shows that the teacher's assessment has been very inaccurate. As a teacher and a researcher, I have often been surprised to find that certain students who seemed sure of themselves were actually anxious about mathematics and vice versa. It seems, therefore, that a part of teachers' evaluation of mathematics students ought to be focused on assessing student attitudes toward mathematics.

Attitudes have been defined in a variety of ways in the psychological literature. Here attitude is used to mean feelings about mathematics and feelings about oneself as a learner of mathematics. This is obviously not a rigorous definition, but it is appropriate for this discussion. Use of this definition is not intended to limit the scope of discussion to general feelings such as liking. More specific feelings, such as confidence and anxiety, are also included in the definition.

The purpose of this chapter is to present knowledge about attitudes toward mathematics, how these attitudes relate to mathematics learning, and to suggest ways in which mathematics teachers can measure these attitudes. In the following sections four specific attitudes that are important in their relationship to students' achievement will be described; confidence in mathematics, mathematics anxiety, attribution of causes of success/failure in mathematics, and the perceived usefulness of mathematics. They have been selected not only because they are important but also because they can be measured easily. Methods of effectively measuring these and other attitudes will be presented.

CONFIDENCE IN MATHEMATICS

Confidence is one of the most important attitudes toward mathematics. Confidence in mathematics has to do with how sure a person is of being able to learn new mathematics, perform well in mathematics class, or perform on mathematics tests. If students are confident, they are apt to learn mathematics better, feel better about themselves, and be more receptive to being taught mathematics. When tasks are approached with confidence, it is more likely that they will be completed successfully than when tasks are begun with a feeling that one will not be able to learn. In addition, people are much more likely to attempt things when they think they have the ability to succeed.

No one knows exactly how confidence is developed. Confidence is related to a student's self-concept, both in general and in relation to the school. Confidence is also affected by the pattern of successes and failures in mathematics that a student experiences throughout school. The confidence level of students in mathematics may be affected by why they think they have succeeded or failed. If success is perceived to have come from some external cause over which one has no control, such as luck, then confidence may not be very strong. Confidence in one's ability to learn mathematics can also be influenced by whether or not one sees mathematics as an appropriate area of study. For example, males who believe that mathematics is a male domain and a highly appropriate area for study tend to feel confident in learning mathematics. Conversely, females who believe that mathematics is a male domain, and therefore an inappropriate area of study for them, may approach mathematics with little confidence. Confidence can easily be linked to the stereotypes of our society.

Research has produced some interesting results about confidence in mathematics and how it affects student achievement. Confidence was the subject of one series of studies dealing with sex-related differences in mathematics achievement in middle and high schools (Fennema and Sherman 1977, 1978; Sherman and Fennema 1977). The main purpose of these studies was to gain insight into sex-related differences in mathematics achievement in grades six through twelve, while controlling for the previous study of mathematics. A second purpose was to gain information about various factors, including confidence, that were hypothesized to be associated with sex-related differences in mathematics achievement.

Students in grades six through twelve were given attitudinal measures (including confidence) and a test of mathematics achievement. The students were from four high schools and the middle schools that fed into these high schools. In both the high school and the middle school grades, when a sex-related difference in achievement in favor of males was found, it was accompanied by a sex-related difference in confidence in favor of males. At both the middle and high school levels, sex-related differences in confidence were found when there was no difference in mathematics achievement. Thus, girls reported confidence in their ability to learn mathematics at lower levels than did boys, both when there might have been some reality basis for this perception, that is, lower achievement, and also when there was no significant difference in achievement.

Moderate relationships between confidence scores and mathematics achievement scores were found in grades six through twelve. Both girls and boys who scored higher in achievement tended to be higher in confidence. In grades nine through eleven the correlations between confidence and mathematics achievement were stronger than for any other attitudinal variables and achievement. In fact, confidence–mathematics-achievement correlations were nearly as high as correlations between spatial visualization, verbal ability, and mathematics achievement.

Of particular interest is the role that scores on the confidence scale played in predicting high school students' plans to continue the study of mathematics. Confidence scores distinguished between students intending and those not intending to take more mathematics courses (Sherman and Fennema 1977). Students with higher levels of confidence planned to take more mathematics than students with lower confidence levels.

Confidence in learning mathematics is significantly correlated with mathematics achievement. Confidence scores are good predictors of which students intend to take more mathematics courses in high school and which do not. Among the affective variables Fennema and Sherman used in their studies, confidence appears to be one of the more important in its relationships to female students' selection of higher-level mathematics courses, and confidence is also important in that it accounted for over one-fifth of the variance in mathematics achievement for females in grades nine to eleven.

Dowling (1978) also studied confidence. She found an even stronger relationship between confidence in mathematics and mathematics achivement than Sherman and Fennema did. Students who had taken more years of high school mathematics expressed more confidence and performed better than students who had taken fewer years of high school mathematics. Sex-related differences in confidence were more prevalent than sex-related differences in achievement.

A recent study (Reyes and Fennema 1980) also focused on confidence in mathematics. In this study sixth-grade students who scored above the mean on a standardized test of mathematics achievement were divided into two groups: students high in confidence and students low in confidence. The high-confidence students and low-confidence students were observed for fifteen to twenty days in their regular mathematics classes to see what types of interactions they had with their mathematics teachers; a variety of characteristics of teacher-student interaction was observed. The initiator of the interaction was recorded as well as the cognitive level of the interaction, that is, whether the interaction was concerned with computation-level tasks or problem-solving-level tasks.

Teachers were found to initiate more interactions with high-confidence students than with low-confidence students; and high-confidence students, in turn, approached the teacher to talk about their mathematics work more frequently than low-confidence students did. A particularly important difference in the patterns of interaction was that high-confidence students tended to have more of the higher cognitive level interactions with teachers than did low-confidence students. Thus, not only do high-confidence students have more interactions with their teachers concerning mathematics, they also have more higher-level interactions with their teachers than low-confidence students do.

It is very important that mathematics teachers spend as much time with students lacking confidence in mathematics as they do with students who are confident in their ability to learn mathematics. By paying less attention to students lower in mathematics confidence, teachers may be unconsciously sending those students the message that they are less capable in mathematics and should therefore expect less of themselves in mathematics.

Confidence in mathematics has been found to be related to student achievement in mathematics and to students' decisions to continue or not continue taking mathematics courses. In terms of sex differences, Dowling as well as Fennema and Sherman found sex-related differences in confidence more frequently than sex-related differences in mathematics achievement. In both the Dowling studies and the Fennema-Sherman studies, in each case where boys and girls differed significantly in mathematics achievement, they also differed significantly in their confidence in mathematics.

Perhaps the most important point for teachers concerning the importance of confidence as an attitude is that student confidence level seems to affect the teacher's treatment of students. Teachers must recognize the importance of treating students low in confidence so that they begin to see their capabilities in mathematics in a realistic way, thus providing equal opportunities for all students. Students who ought to take three or four years of mathematics in high school but lack the confidence to realize this should be encouraged by their teachers. Teachers must be extremely careful in examining the way they interact with high- and low-confidence students. The tendency is for the well-behaved, hard-working, low-confidence students to receive less attention than they ought to from the teacher. It is essential for teachers to know which of their students are high or low in confidence and to understand the effect this can have on student performance.

MATHEMATICS ANXIETY

Interest in the causes of and remedies for mathematics anxiety has greatly increased over the past few years. Mathematics-anxiety clinics for college and high school students are found in increasing frequency in many areas of the country. Mathematics anxiety, which is similar to lack of confidence in mathematics, has been found to be related to both mathematics achievement and the election of mathematics courses.

General anxiety has been a topic of study for many years. Psychologists have studied anxiety extensively and have used a variety of definitions and theories to describe and explain anxiety. The literature growing out of this intensive study leads one to believe that caution must be exercised in viewing anxiety as an illness. Anxiety is not a pathological state that should be cured. Rather, anxiety is an essential and complex feeling that does not always function in a totally facilitating manner (Sieber, O'Neil, and Tobias 1977). Anxiety may, however, be helpful or facilitative in learning and performing simple tasks. In contrast, anxiety is often detrimental and debilitating to performance on complex tasks.

One of the ways in which anxiety is understood to stand in the way of performance, particularly in evaluative situations, is that anxiety tends to take the learner's attention away from the task at hand (for example, a test question) and focus that attention internally (I. Sarason 1975). Anxious students often spend time engaged in negative self-talk, such as "I am stupid" or "I know I can't learn math." In a learning or evaluative situation, it is reasonable that placing one's attention on such negative self-thoughts will interfere with attention to the task at hand and thereby reduce the level of performance on that task.

Even though extensive work has been done on anxiety, comparatively little research on specific mathematics anxiety has been reported. There seems to be some connection between mathematics anxiety and general anxiety; however, this relationship has not been studied in depth. It has been demonstrated that "mathematics anxiety exists among many individuals who do not ordinarily suffer from any other tensions" (Richardson and Suinn 1972, p. 551). So even though mathematics anxiety is related to other types of anxiety, many capable people who are not generally anxious are anxious about mathematics.

Mathematics anxiety can be described as involving feelings of tension and anxiety that interfere with the manipulation of numbers and the solving of mathematical problems in a wide variety of ordinary life and/or learning situations. Mathematics anxiety can prevent students from doing their best, from passing fundamental mathematics courses, or from pursuing advanced courses in mathematics or the sciences.

> Among nonstudents, mathematics anxiety may be a contributor to tensions during routine or everyday activities such as handling money, balancing bank accounts, evaluating sales prices, or dividing work loads. (Richardson and Suinn 1972, p. 551)

There are some general conclusions about mathematics anxiety that can be drawn, and some representative results of research dealing specifically with mathematics anxiety help to understand these conclusions. A study by Betz (1978) examined the patterns of occurrence of mathematics anxiety among three groups of freshmen and sophomore college students. One group (Math 1) consisted of students in a low-level mathematics course designed to review high school algebra. A second group (Math 2) came from a more advanced course, precalculus, for students planning majors in mathematics, engineering, premedicine, and the physical sciences. The third group (Psych I) consisted of students from an introductory psychology course. The results suggest that mathematics anxiety is a problem for many college students, including many in advanced mathematics classes whose majors require an extensive background in mathematics. In the low-level classes, women were significantly more anxious about mathematics than men were. The results also suggest that the number of years of high school mathematics taken "strongly influences how a college student will feel about math" (Betz 1978, p. 446). The level of mathematics anxiety reported was related to scores on a standardized mathematics achievement test. People with high achievement scores tended to report low mathematics anxiety.

Females have tended to report higher levels of mathematics anxiety than males have (Betz 1978, Perl 1979). Females have also reported higher levels of other types of anxiety than males. It is difficult to separate these results from the known tendency of females to be more willing to report their feelings than males are. Females typically are more open about emotions than males, and some researchers believe that higher anxiety reported by women is an artifact of this (Maccoby and Jacklin 1974). Researchers have not yet determined whether the differences in reported anxiety are due to true differences in anxiety or not. Care must be taken not to add mathematics anxiety to the stereotype of females as being less capable in mathematics than males. Mathematics anxiety occurs in both females and males. Not all females are anxious nor are all males confident. As teachers become aware of the problems of the anxious student, importance should not be placed on who is anxious but on the causes, effects, and remedies of mathematics anxiety.

A consistent negative relationship has been found between anxiety and mathematics achievement such that high achievement is related to low anxiety for students from grade school through college (Aiken 1970a, 1970b, 1976; Betz 1978; Callahan and Glennon 1975; Crosswhite 1972; S. Sarason et al. 1960; Szetela 1973). None of these studies, however, has demonstrated a clear cause-effect relationship between mathematics anxiety and achievement in mathematics. Szetela (1970, 1973) found that as anxiety about mathematics tests increased, performance of eighth graders in mathematics decreased. Controlling for intelligence, Szetela found some, though not conclusive, evidence that mathematics test anxiety is consistently related to mathematics achievement, but no clear replicated evidence shows that high levels of mathematics anxiety bring about decreased achievement in mathematics.

Much of the current interest in mathematics anxiety is concerned with reducing anxiety among mathematics students. A variety of interventions designed to reduce mathematics anxiety have been developed and implemented. Most of these programs have been used with college students and adults not currently in school. Some programs have been aimed particularly at mathematics-anxious women. Few interventions have been carefully evaluated. This is unfortunate. Some programs designed to reduce mathematics anxiety claim not only to reduce anxiety but also to improve mathematics achievement and increase election of mathematics courses. A careful evaluation of the effects of a program should be carried out in terms of changes in anxiety, achievement, and course election brought about by the intervention program. Some programs have been shown effective in reducing levels of anxiety, but very few have been effective in improving achievement scores or increasing mathematics course election.

There are several types of mathematics anxiety interventions. Some focus on the content of mathematics. They work on the assumption that mathematics anxiety is caused by a lack of understanding of mathematics, and they attempt to increase knowledge of mathematical concepts and thereby reduce mathematics anxiety. One such intervention for which there are evaluative data was developed and implemented by Crumpton (1977). She used the Mathematics Anxiety Rating Scale (MARS) to determine content areas in mathematics that created anxiety among a group of college students of mathematics. Through the use of a mathematics achievement test, content areas were identified for which students' competency was low. Instruction was then carried out in the high-anxiety–low-competency content areas. No discussion of mathematics anxiety was included in this treatment. Results showed that increased competence in mathematics was accompanied by reduced mathematics anxiety.

Another type of mathematics anxiety reduction program focuses on teaching students to deal with anxiety effectively and thus reduce the effects of mathematics anxiety on performance in the subject. One of the interventions that has been effective in reducing mathematics anxiety is desensitization behavior therapy (Hart 1977; Richardson and Suinn 1972). Desensitization has been used successfully in reducing other specific anxieties such as test anxiety as well (Aiken 1970a, 1976; Sieber, O'Neil, and Tobias 1977; Wine 1971). The theory of systematic desensitization consists of relaxing the mathematics-anxious individual, usually through a procedure of deep muscle relaxation. Once in a relaxed state, the individual is directed to picture a sequence of scenes. The early scenes are only slightly anxiety arousing. After each scene the individual is directed back into that original relaxed state. The scenes become progressively more anxiety arousing, but the individual is maintained in a relaxed state. In this way the anxious person learns to associate mathematics with a pleasant, relaxed feeling. Length of treatment varies greatly from single one-to-two-hour sessions to several such sessions. Systematic desensitization is probably best administered by a trained counselor or therapist rather than the mathematics teacher, since classroom teachers normally do not have the necessary counseling background.

A third approach to the reduction of mathematics anxiety uses a combination of the first two approaches. The student who is anxious about mathematics is given instruction in mathematics and also counseling in reducing mathematics anxiety. The Tobias (1978) mathematics anxiety clinic is of this type, as is the Hendel and Davis (1978) intervention. In this combination approach, a mathematics instructor often works together with a counselor. The counselor helps the mathematics teacher avoid instructional techniques that arouse anxiety, while the teacher helps to clear up misconceptions about mathematics content during counseling sessions.

There is little research on the relative effectiveness of these three approaches on mathematics anxiety reduction. In fact one of the weaknesses of many mathematics anxiety treatments has been the absence of a well-done evaluation of effectiveness.

One of the most publicized mathematics anxiety reduction programs was developed at Wesleyan University by Sheila Tobias, Bonnie Donady, and Susan Auslander. Tobias (1978) describes their program along with a detailed description of her conception of mathematics in a book, *Overcoming Math Anxiety*. This book, along with *Mind Over Math* by Stanley Kogelman and Joseph Warren (1978) would be appropriate for adults and high school students who are "math anxious." Another good resource for possible ways of dealing with mathematics anxiety is a report of the Proceedings of the Conference on the Problem of Math Anxiety (1978). A variety of interventions are presented that may be helpful in designing new programs to reduce mathematics anxiety.

Mathematics anxiety has been found to be strongly related to mathematics confidence (Fennema and Sherman 1976). Students high in mathematics confidence tend to be low in mathematics anxiety and vice versa. Since the two attitudes are closely linked, it makes sense to think of them as one attitude with confidence as its positive manifestation and anxiety its negative one. Teachers should take care not to focus only on anxiety.

What does all this mean for teachers? It does not mean that mathematics class time should be entirely converted into a therapy session for anxious students. It does mean, however, that teachers should be aware of certain students who are capable of performing well in mathematics but may be blocked by mathematics anxiety.

Objective measurement of anxiety should be done to supplement teacher observations. Students often learn to mask their feelings from adults and even other students. An objective measurement added to what a teacher observes will reveal many students who are anxious. Teachers should take an objective look at the environment in their classes. Are certain approaches to teaching mathematics more anxiety arousing than others? This is not to imply that nothing which triggers anxiety should occur in classrooms. Teachers should be aware, however, of how students react to the class environment and procedures, which students are anxious, and what situations trigger that anxiety.

Teachers can also make help available to the highly mathematics anxious by working with school counselors.

ATTRIBUTION THEORY

Attribution theory has recently gained attention from educators and appears to have certain logical relevance to mathematics. Attribution theory deals with an individual's beliefs about the causes of behavior, in particular the reasons that one succeeds or fails in a variety of experiences.

A person can attribute success or failure to a variety of causes. For example, students who get A's on a mathematics test may attribute that success to their ability, to having worked hard, to a good textbook, to help received from the teacher or a friend, to luck, or to the fact that the test was an easy one. Perceived causes of success/failure or attributions fall into four major categories: ability, effort, task difficulty, and luck. Weiner (1972) organized these four categories along two dimensions, locus of causality and stability (Table 10-1).

The internal/external or locus of causality dimension indicates whether the cause is controlled within the person or outside the person. Ability is internal since it is determined within the individual. Whether or not a student decides to pay attention and work in mathematics class is internal or within that person's control. External causes of success or failure are things such as the quality of the explanations the teacher gives or the difficulty of the problems on tomorrow's assignment. The stable/unstable dimension has to do with whether or not the factor may vary from time to time. Effort or luck as a cause of success/failure may vary from situation to situation, while ability and

Table 10-1

Category scheme for perceived causes of success/failure

STABILITY	LOCUS OF CAUSALITY	
	Internal	External
Stable	ability	task difficulty
Unstable	effort	luck

task difficulty remain constant. As shown in Table 10-1, ability is stable and internal, effort is unstable and internal, task difficulty is stable and external, and luck is unstable and external.

Research yields patterns of attribution that are related to characteristics of people and are relatively consistent within individuals. People high in self-concept tend to attribute positive outcomes internally and negative outcomes externally. In contrast, people with low self-concepts more often ascribe positive outcomes to external causes, while tending to attribute negative outcomes internally (Ickes and Layden 1978).

Males and females on the average seem to differ in their patterns of attribution. A very consistent finding (Deaux 1976) is that females hold lower expectations of their own performance than males do of theirs. Females tend to attribute their own successes to unstable external causes and their failures to internal stable causes. In contrast, males tend to attribute their own successes to ability and to search for unstable external causes to explain failure.

The notion of expectancies seems to be critical in determining attributions (Deaux 1976). "Subjects who approached a task with a high degree of confidence attributed failure externally and success internally, while these trends were reversed for subjects initially low in confidence" (Deaux 1976, p. 346). In addition, sex differences in attribution seem only to appear when expectations by females and males differ (McMahan 1973).

If these patterns of expectancies and attributions generalize to the learning of mathematics, it may be that females (and males) are caught in a cycle. Female students in grades six to twelve tend to be less confident in mathematics than their male counterparts. This lower confidence leads to more internal attribution of failure and external attribution of success. Attribution of success to luck or to an easy task provides no foundation for increasing confidence or expectancies in the future.

It is not known that lower confidence in one's ability to learn and perform in mathematics causes lower achievement in mathematics. There is evidence that: (a) confidence is more strongly related to mathematics achievement than some other affective variables (Fennema and Sherman 1977, 1978), and (b) confidence is a significant factor that distinguishes between students who intend to enroll in another year of mathematics and those who do not (Sherman and Fennema 1977). Thus, a self-perpetuating cycle of lower confidence, internal attribution of failure, and external attribution of success may exist, which in turn can affect achievement or decisions to take more mathematics. The existence of this cycle is hypothesized. More study is needed to see if such a cycle actually exists for female or low-confidence mathematics students. A more detailed view of attribution and mathematics education is presented by Fennema (in press).

Relationships between achievement-related behaviors such as persistence and patterns of attribution have been studied by social psychologists in laboratory settings. When failure is attributed to stable causes (ability and/or task difficulty), subjects tend to persist less. Failure tends to be met with greater persistence when the cause of failure is seen as some unstable cause like bad luck or lack of effort (Weiner et al. 1972).

Bar-Tal (1978) indicates that individuals who are high in achievement motivation tend to select achievement-related activities more frequently than individuals low in achievement motivation. He also found consistent variation in the attributional styles of people who differ in the need to achieve. Those high in the need to achieve tend to attribute success to internal causes (ability and effort). Individuals low in achievement need tend to attribute their success to external causes and not to effort or ability.

Care must be taken in making direct implications from attribution theory and research. People may differ some in their patterns of attributions, but this does not necessarily lead to the logical conclusions that: (a) there is something wrong with certain attributional patterns, or (b) we should attempt to change patterns of attribution.

Though student behavior in the classroom may differ from behavior in a laboratory setting, teachers should be aware of students' attributions of success and failure. It is interesting to begin paying attention to the way we as individuals attribute causes to our successes and failures. We may also want to monitor the attributions we make for students' successes and failures. Do we assume that certain factors cause the success or failure of our students? Are our attributions accurate? Do certain of our students attribute their successes to luck and their failures to lack of ability? An awareness of attribution patterns can be fascinating, but more they may be important in determining the amount of persistence and confidence with which students approach mathematics. It certainly seems logical that people should feel they have some control over their lives and their learning of mathematics. If students believe that effort is a cause of success and failure, perhaps they will also see that they can have some control over the mathematics they learn by controlling the amount of effort they exert.

PERCEIVED USEFULNESS OF MATHEMATICS

Students vary in how useful they view mathematics to be, both for their current needs and for the future. This perceived usefulness of mathematics is an important factor in determining whether students will elect to take mathematics classes. In addition, it probably affects how much effort they will actually expend in the mathematics classes. This factor becomes increasingly important as girls and boys progress through middle school into high school. Some students enjoy learning and studying mathematics, while others do not. Many of the latter students elect to take three or four years of mathematics in high school, even if they do not particularly like it, if they know that mathematics is necessary for their career goals. However, without specific career goals or information about the level of mathematical knowledge necessary for reaching these goals, other students drop out. For this reason as well as others, many students—including disproportionately large numbers of female and minority students—take only the minimum mathematics required in high school. By doing so, they narrow their options for the future. A better understanding of the importance of mathematics in a wide range of careers and post high school education is important for students as they make decisions about how much mathematics to take in high school. For example, teachers and their students should be aware that the following careers require at least three years of high school mathematics: landscape architecture, anthropology, business administration, medical technology, nursing, and drafting. An excellent resource on the mathematics required for entrance into a variety of majors at both colleges and vocational schools is provided in materials developed by Fennema et al. (1979) under a grant from the United States Office of Education.

Two sets of studies in particular have examined students' views on the usefulness of mathematics (Perl 1979; Fennema and Sherman 1977, 1978). Fennema and Sherman found that among middle school and high school students, those who receive higher scores in mathematics achievement see mathematics as more useful than the lower achievers do. They also demonstrated that high school students' views about usefulness of mathematics do affect their election of mathematics courses. Those who perceive mathematics as useful to them tend to elect more mathematics courses.

Perl (1979) analyzed some of the data from grades ten through twelve from the National Longitudinal Study of Mathematical Abilities. She found that views on the usefulness of mathematics discriminated between students who elect and students who do not elect to take more mathematics courses. Perl also studied influences on differential course election by girls and boys. She identified perceived usefulness as the most important attitudinal factor in explaining the differences in mathematics course election between girls and boys. Similar conclusions concerning usefulness were reached by Haven (1971) and Hilton and Berglund (1974) in their studies of high school students. Several studies have found that boys perceive mathematics to be more useful to them than girls do (Haven 1971; Hilton and Berglund 1974; Perl 1979). Fennema and Sherman (1977, 1978) found sex-related differences in students' views of the usefulness of mathematics only in those schools where they also found differences in mathematics achievement between girls and boys.

The simplistic truth about this attitude appears to be overlooked as teachers, parents, and counselors help prepare students for their future lives. Teachers are in a good position both to assess how useful their students view mathematics to be and to give students information about the importance of studying mathematics. Of the four specific attitudes discussed in this chapter, usefulness may be the easiest attitude to change, and teachers are in the best position to bring about change in students' views of the usefulness of mathematics. Teachers spend more school hours with students than any other adults and can easily give students information about the number of years of mathematics required for entrance into a variety of careers as well as those needed to keep options open for the future. It is clear that the perceived usefulness of mathematics is an important affective variable for teachers to understand and measure.

MEASUREMENT OF ATTITUDES

Early interest in measuring attitudes toward mathematics focused mainly on global attitudes that are not particularly useful. The first paper-and-pencil instruments to determine attitudes toward mathematics generally measured student like or dislike of the subject. One of the instruments, published by Dutton in 1951, defined attitudes as "the emotionalized feelings of students for or against something" (1951, p. 84). In other words, the scale measured whether or not students liked mathematics. One of the more commonly used scales is the Mathematics Attitude Scale (Aiken and Dreger 1961), which was designed to measure general attitudes toward mathematics. The items of the Mathematics Attitude Scale cover a variety of affective aspects, such as liking/disliking and anxiety. Use of this scale may not be of much help to teachers, however, since what it measures is not well defined.

The scales from the National Longitudinal Study of Mathematical Abilities (NLSMA) are an improvement over earlier ones because attitudes are defined more precisely. Instead of being conceptualized as a single dimension, attitudes were divided into specific categories (for example, math easy versus hard and math fun versus dull). Self-concept and anxiety scales were specifically included. This seems to be the first time that personality factors were included as defined dimensions of the attitudinal domain in the literature of mathematics education.

Since NLSMA, other multidimensional scales have been developed to measure a variety of specific well-defined attitudes, such as enjoyment of mathematics, confidence in learning mathematics, mathematics anxiety, and the value of mathematics (Bowling, 1976; Fennema and Sherman 1976; and Sandman 1973). The Fennema-Sherman Mathematics Attitude Scales are an example of multidimensional scales. This set of nine scales includes measures of confidence, anxiety, motivation, usefulness, attitude toward success, the stereotyping of mathematics as a male domain, students' perceptions of their teachers' attitudes toward them as learners of mathematics, students' perceptions of their mothers' attitudes toward them as learners of mathematics, and students' perceptions of their fathers' attitudes toward them as learners of mathematics. Each scale contains twelve statements, half of which are stated positively ("I like math puzzles") and half are stated negatively ("A math test would scare me"). Students respond to each statement by selecting one of five alternatives: strongly agree, agree, undecided, disagree, and strongly disagree. These scales are appropriate for use with middle school and high school students. Scores on these scales do not represent absolute levels of a particular attitude, and no norms are available. The scores do indicate relative levels of attitudes among students, however, and are useful to assess change in attitudes.

Refinement beyond the multidimensional scale has recently been achieved not only by looking at specific aspects of attitudes but also by studying attitudes about specific aspects of mathematics. An example of such a scale is the Mathematics Confidence Scale (MCS) developed by Dowling (1978). Dowling took a specific attitude, confidence in mathematics, and assessed student confidence for three dimensions of mathematics tasks: three components of content (arithmetic, algebra, and geometry), three levels of cognitive demand (computation, comprehension, and application), and two problem contexts (real settings and abstract settings).

The MCS has a somewhat unusual format for an attitude scale. In Part I, students are given eighteen problems representative of the three dimensions of mathematics tasks. For Part I, students do *not* work the eighteen problems. They use a five-point scale ranging from "no confidence at all" to "complete confidence" to indicate their level of confidence in their ability to solve the problems. In Part II, students are asked to solve problems similar to those in Part I. In this way the MCS measures confidence about particular problems and also actual student performance on similar mathematical tasks.

While the MCS was designed primarily for use with college students, it or an adaptation, made by adjusting the level of the problems used, would be useful at other levels. The only limitation would be in students' ability to read and their ability to understand and respond about their confidence level. This makes the MCS format especially useful for teachers.

Another example of a scale that measures one attitude but divides mathematics into parts is the Mathematics Anxiety Rating Scale (MARS) (Suinn 1970). It was developed and normed with adults and has both students and nonstudents as its intended users. The scale consists of ninety-eight items describing situations that could arouse mathematics anxiety. Respondents indicate how much they are frightened by the described situations by checking one of the following: not at all, a little, a fair amount, much, or very much. The items vary widely. Some of the items seem capable of arousing little anxiety ("reading an historical novel with many dates in it"), while others could conceivably be quite fear producing ("being given a 'pop' quiz in a math class"). The items represent many different situations, some from school settings and others from settings where mathematics might be needed in a business or social framework. Each item asks the level of fearfulness the setting arouses in the individual at the present time.

There are a variety of techniques for measuring attitudes toward mathematics: questionnaires, interviews, behavior observation, and paper-and-pencil scales like the ones already discussed. The Likert scale, a particular type of paper-and-pencil scale, is a frequently used method of measuring attitudes toward mathematics. The Fennema-Sherman scales are examples of the Likert Scale, since subjects respond by marking their level of agreement/disagreement with each statement in the scale. As an example of a Likert scale, the Fennema-Sherman scale Confidence in Learning Mathematics is shown in Figure 10-2 with sample responses marked. This scale is designed:

> to measure confidence in one's ability to learn and to perform well on mathematical tasks. The dimension ranges from distinct lack of confidence to definite confidence. The scale is not intended to measure anxiety and/or mental confusion, interest, or zest in problem solving. (Fennema and Sherman 1976, p. 4)

To score a Likert scale, a number from one to five is first assigned for each item. Five indicates most confidence, while one indicates least confidence. So a positively stated item marked "strongly agree," as item 3 is, would receive a score of five. Item 4 would receive two points since it is a positive statement marked "disagree." Item 9 receives five points since it is a negatively stated item marked "strongly disagree." A negatively stated item marked with an "agree" response (item 11) receives two points. The score for the scale is obtained by simply adding the scores of the individual items. In the example in Figure 10-2, the sum of the item scores is forty-five. Since there are no norms for this scale, the score of forty-five is only meaningful when compared to scores of other people or to scores of the same person obtained at different times. The range of possible scores is from a low score of twelve to a high score of sixty, with thirty-six as the midpoint.

If the Confidence in Learning Mathematics scale were to be given to a class, it would be presented somewhat differently. The weights of the items would be omitted and the positive and negative items would be randomly ordered. Instructions would be given to students about how to mark the items. A sample item or two is essential to be sure students understand the procedure. If students are being asked to respond to more than one of these scales, the items of both scales should be randomly ordered.

Most of the scales described in this chapter were developed for use in research. Teachers who are not necessarily interested in research might still find these scales helpful for use in their classrooms. Teachers could also devise their own Likert scales. The first step in writing a scale is to decide exactly what the scale ought to measure. This description is essential, should be carefully considered, and is probably

*Weight**			*Points*
+	1. Generally I have felt secure about attempting mathematics.	SA (A) U D SD	4
+	2. I am sure I could do advanced work in mathematics.	SA A (U) D SD	3
+	3. I am sure that I can learn mathematics.	(SA) A U D SD	5
+	4. I think I could handle more difficult mathematics.	SA A U (D) SD	2
+	5. I can get good grades in mathematics.	SA (A) U D SD	4
+	6. I have a lot of self-confidence when it comes to math.	SA (A) U D SD	4
–	7. I'm not good in math.	SA A U (D) SD	4
–	8. I don't think I could do advanced mathematics.	SA A U (D) SD	4
–	9. I'm not the type to do well in math.	SA A U D (SD)	5
–	10. For some reason, even though I study, math seems unusually hard to me.	SA A U (D) SD	4
–	11. Most subjects I can handle easy, but I have a knack for flubbing up math.	SA (A) U D SD	2
–	12. Math has been my worst subject.	SA A U (D) SD	4

*Weights and points omitted when the scale is given to students.

Figure 10-2

Fennema-Sherman scale: confidence in learning mathematics

best put in writing. Then the scale developer should write as many items as possible that fit the description. After all of the items have been written, the best ten to twenty items are selected for use in the scale. The best items may be selected by trying all of the items with a few students and/or by rereading the items carefully and critically. Though this procedure of scale development would not be adequate for research purposes, it certainly would give the teacher some information about student attitudes.

Once an attitude scale has been scored, whether it is a published scale or one written by the teacher, the most important task is to decide what meaning the scores have, because scores on an attitude scale do not necessarily mean anything. The information obtained from administering an attitude scale is no better than the items contained in the scale. For example, if a scale is titled "Mathematics Anxiety" but its items are nearly all concerned with whether or not mathematics is useful in everyday life, a score on that scale is not a valid measure of mathematics anxiety. The best way to know exactly what a scale measures is to examine each item carefully. By taking the composite of the items in the scale, one can determine what the numerical score for the scale indicates.

In the past mathematics teachers have spent little time objectively measuring student attitudes toward mathematics. Yet measuring student levels of attitudes such as confidence, anxiety, usefulness, and attribution of success/failure can give teachers important information. Such information can help them improve the quality of their decisions about teaching mathematics. Teachers' improvement in their understanding of student attitudes toward mathematics can contribute to increasing the number of students who enroll in high school mathematics courses. Knowledge about students' attitudes toward mathematics will also enable teachers to teach more effectively and successfully.

REFERENCES

Aiken, Lewis R., Jr. "Attitudes toward Mathematics." *Review of Educational Research* 40 (October 1970): 551–96(a).

Aiken, Lewis R., Jr. "Nonintellective Variables and Mathematics Achievement: Directions for Research." *Journal of School Psychology* 8, no. 1 (1970): 28–36(b).

Aiken, Lewis R., Jr. "Update on Attitudes and Other Affective Variables in Learning Mathematics." *Review of Educational Research* 46 (Spring 1976): 293–311.

Aiken, Lewis R., Jr., and Dreger, Ralph M. "The Effect of Attitudes on Performance in Mathematics." *Journal of Educational Psychology* 52 (February 1961): 19–24.

Bar-Tal, Daniel. "Attributional Analysis of Achievement-Related Behavior." *Review of Educational Research* 48 (Spring 1978): 259–71.

Betz, Nancy E. "Prevalence, Distribution, and Correlates of Math Anxiety in College Students." *Journal of Counseling Psychology* 25 (September 1978): 441–48.

Bowling, J. Michael. "Three Scales of Attitude toward Mathematics." Doct. diss., Ohio State University, 1976.

Callahan, Leroy G., and Glennon, Vincent J. *Elementary School Mathematics: A Guide to Current Research.* Washington, D.C.: Association for Supervision and Curriculum Development, 1975.

Crosswhite, F. Joe. *Correlates of Attitudes toward Mathematics*, National Longitudinal Study of Mathematical Abilities, Report No. 20. Palo Alto, Calif.: Stanford University Press, 1972.

Crumpton, Sharon D. "A Mathematical Anxiety Reduction Project." Doct. diss., University of Tennessee, 1977.

Deaux, Kay. "Sex: A Perspective on the Attribution Process." In *New Directions in Attribution Research*, vol. 1, edited by John H. Harvey, William J. Ickes, and Robert F. Kidd. Hillsdale, N.J.: Lawrence Erlbaum Associates, 1976, pp. 335–52.

Dowling, Delia Mary "The Development of a Mathematics Confidence Scale and Its Application in the Study of Confidence in Women College Students." Doct. diss., Ohio State University, 1978.

Dutton, Wilbur H. "Attitudes of Prospective Teachers toward Arithmetic," *Elementary School Journal* 52 (October 1951): 84–90.

Fennema, Elizabeth. "Attribution Theory and Achievement in Mathematics." In *The Growth of Insight in Children's Thinking*, edited by Steven Yussen, in press.

Fennema, Elizabeth; Becker, Ann D.; Wolleat, Patricia L.; Pedro, Joan Daniels. *Multiplying Options and Subtracting Bias: An Intervention Program Designed to Eliminate Sexism from Mathematics Education.* Washingon, D.C.: U.S. Department of Health, Education, and Welfare, Office of Education, 1979.

Fennema, Elizabeth, and Sherman, Julia. "Fennema-Sherman Mathematics Attitude Scales." *JSAS Catalogue of Selected Documents in Psychology* 6 (1976): 31 (Ms. No. 1225).

Fennema, Elizabeth and Sherman, Julia. "Sex-Related Differences in Mathematics Achievement, Spatial Visualization and Affective Factors. *American Educational Research Journal* 14 (1977) 51–71.

Fennema, Elizabeth, and Sherman, Julia. "Sex-Related Differences in Mathematics Achievement and Related Factors: A Further Study." *Journal for Research in Mathematics Education* 9 (May 1978): 189–203.

Hart, Laurie E. "The Effects of a Form of Systematic Desensitization on Test-Anxious Mathematics Students." Master's thesis, University of Texas at Austin, 1977.

Haven, Elizabeth W. "Factors Associated with the Selection of Advanced Academic Mathematics Courses by Girls in High School." Doct. diss., University of Pennsylvania, 1971.

Hendel, Darwin D., and Davis, Sandra O. "Effectiveness of an Intervention Strategy for Reducing Mathematics Anxiety." *Journal of Counseling Psychology* 25 (September 1978): 429–34.

Hilton, Thomas L., and Berglund, Gosta W. "Sex Differences in Mathematics Achievement—A Longitudinal Study." *Journal of Educational Research* 67 (January 1974): 231–37.

Ickes, William J., and Layden, Mary Ann. "Attributional Styles." In *New Directions in Attribution Research*, vol. 2, edited by John H. Harvey, William J. Ickes, and Robert F. Kidd. Hillsdale, N.J.: Lawrence Erlbaum Associates, 1978, pp. 119–52.

Kogelman, Stanley, and Warren, Joseph. *Mind over Math.* New York: Dial Press, 1978.

Maccoby, Eleanor E., and Jacklin, Carol N. *The Psychology of Sex Differences.* Stanford, Calif.: Stanford University Press, 1974.

McMahan, Ian D. "Relationships between Causal Attributions and Expectancy of Success." *Journal of Personality and Social Psychology* 28 (October 1973): 108–14.

Perl, Teri H. "Discriminating Factors and Sex Differences in Electing Mathematics." Doct. diss., Stanford University, 1979.

Proceedings of the Conference on the Problem of Math Anxiety. Fresno, Calif.: School of Natural Sciences, California State University, Fresno, 1978.

Reyes, Laurie H., and Fennema, Elizabeth. "Teacher/Pupil Influences on Sex Differences in Mathematics Confidence." In preparation, 1980.

Richardson, Frank C., and Suinn, Richard M. "The Mathematics Anxiety Rating Scale: Psychometric Data." *Journal of Counseling Psychology* 19 (November 1972): 551–54.

Sandman, Richard S. "The Development, Validation, and Application of a Multidimensional Mathematics Attitude Instrument." Doct. diss., University of Minnesota, 1973.

Sarason, Irwin G. "Anxiety and Self-Preoccupation." In *Stress and Anxiety*, vol. 2, edited by Irwin G. Sarason and Charles D. Spielberger. Washington, D.C.: Hemisphere Publishing Corp., 1975, pp. 27–44.

Sarason, Seymour B.; Davidson, Kenneth S; Lighthall, Frederick F.; Waite, Richard R.; Ruebush, Britton K. *Anxiety in Elementary School Children.* New York: Wiley, 1960.

Sherman, Julia, and Fennema, Elizabeth. "The Study of Mathematics by High School Girls and Boys: Related Variables." *American Educational Research Journal* 14 (Spring 1977): 159–68.

Sieber, Joan E.; O'Neil, Harold F., Jr.; and Tobias, Sigmund. *Anxiety, Learning, and Instruction.* Hillsdale, N.J.: Lawrence Erlbaum Associates, 1977.

Suinn, Richard M. "The Mathematics Anxiety Rating Scale." (Unpublished manuscript, Department of Psychology, Colorado State University, Fort Collins, Colo., 1970.

Szetela, Walter. "The Effects of Test Anxiety and Success-Failure on Mathemetics Performance in Grade Eight." Doct. diss., University of Georgia, 1970.

Szetela, Walter. "The Effects of Test Anxiety and Success/Failure on Mathematics Performance in Grade Eight." *Journal for Research in Mathematics Education* 4 (May 1973): 152–60.

Tobias, Sheila. *Overcoming Math Anxiety.* New York: W. W. Norton and Co., 1978.

Weiner, Bernard. "Attribution Theory, Achievement Motivation, and the Educational Process." *Review of Educational Research* 42 (Spring 1972): 203–15.

Weiner, Bernard; Heckhausen, Heinz; Meyer, Wulf-Uwe; Cook, Ruth E. "Causal Ascriptions and Achievement Behavior: A Conceptual Analysis of Effort and Reanalysis of Locus of Control." *Journal of Personality and Social Psychology* 21 (February 1972): 239–48.

Wine, Jeri. "Test Anxiety and Direction of Attention." *Psychological Bulletin* 76 (August 1971): 92–104.

11. Sex-Related Issues in Mathematics Education

Ann K. Schonberger

In the past decade national and international attention has been focused on improving the status of women. In this country, women's demands for equality have led to identification of a number of target areas in which women's roles have been limited in the past; primary among these are economic and intellectual activities. In the economic sphere, men far outnumber women in the scientific and technical job market, occupations that require a high level of mathematical competence. Although other causes for this imbalance have been hypothesized, male-female differences in mathematical ability are often suggested as reasons (Carnegie Commission 1973). Furthermore, as women become more independent in buying and maintaining homes, cars, and other products, mathematics becomes more important in their everyday lives. In the intellectual sphere, mathematics has also increased in importance both because of its use as one index of general scholastic ability in post-secondary-school admissions and because of the trends toward quantification in social sciences and even in the humanities. To quote a noted mathematics educator: "our instruction serves to develop the capacity of the human mind for the observation, selection, generalization, abstraction, and construction of models for use in solving problems in other disciplines" (Fehr 1974, p. 27).

PERFORMANCE COMPARISONS

Certainly equal participation by women in our society's economic and intellectual activities requires mathematical competence equal to that of men. Are there, in fact, sex-related differences in mathematics achievement appearing in elementary or secondary school? That question can be investigated because research on mathematics achievement has often used sex as a variable, and there are a number of summaries of research on sex-related differences in this area. As summarized by Fennema (1978), reviews published before 1974 usually conclude that male superiority in mathematics achievement was evident by adolescence, if not before. This position was also taken by Maccoby and Jacklin (1974) in their often-cited review, although their summary has been criticized by Sherman (1978). Fennema's (1974) review, however, indicates that the only sex-related differences apparent were among some groups of high school students and only on higher-level cognitive tasks such as problem solving; better performance by females was found occasionally on computational tasks. My own review (Schonberger 1978) of the research on problem-solving available in 1975 suggests that better male performance, if found at all, was usually limited to students of higher ability and to certain types of problems. More recently completed research (Fennema and Sherman 1977, 1978; Schonberger 1978) found few sex-related differences, even in problem solving.

These discrepancies between earlier and later studies can be attributed to various factors, a discussion of which can help readers evaluate other research on sex-related differences they may encounter. One factor is the possibility that students have changed in the twenty years covered by this research. Certainly these have been years of great change for youth in general and women in particular. Evidence from the Project Talent follow-up (Flanagan 1976) indicates that this may be part, but not all, of the answer.

Another important source of the disparate findings between earlier and later studies is the content of the tests. In the National Longitudinal Study of Mathematical Abilities done in the mid-1960s, using students in grades four through eleven, there was a statistically significant difference in performance favoring males on almost all of the number series scales at the highest cognitive level (data summarized in Schonberger 1978). Close examination revealed that these scales contained problems about people, and in virtually *all* cases in which the sex of a person was specified, the person was male. More recent studies of mathematical problem solving using tests constructed with an effort to eliminate sex bias have found few, if any, sex-related differences in grades four through eight (Fennema and Sherman 1978; Schonberger 1978).

A large-scale study by researchers at the Educational Testing Service (Donlon, Ekstrom, and Lockheed 1979) also indicates that this discrepancy is not merely due to the change in student population. They evaluated items and performance data from grade ten of the Sequential Tests of Educational Progress (STEP) and from grades five and eight of the Iowa Tests of Basic Skills (ITBS). On the people-centered items of the mathematics tests, the only significant differences were in favor of males in grade ten on items with male references or roles on the STEP and in favor of females in grade five on items with female or neutral references, actors, or roles on the ITBS. Donlon (1973) also analyzed the data from the May 1964 administration of the Scholastic Aptitude Test—Mathematics, the items of which were categorized as algebra, geometry, and subject matter. He indicated that the forty-point differences between average scores for males and females could grow to sixty points if the number of subject-matter items (in this test, male dominated) were increased; he suggested that the difference could decrease by twenty points if only algebra items were used.

Donlon's paper raised another issue, which is probably the most important in evaluating research on sex-related differences: controlling for opportunity to learn or the number and type of mathematics courses taken. Of the seniors taking College Board Examinations in 1971–72, half of the males had taken four years of mathematics whereas only a third of the females had done so. Some of the large-scale studies from the 1960s that were cited in the earlier reviews can be faulted for this lack of control. For example, the Project Talent Study (Flanagan et al. 1964) reported only slight differences in grade twelve. A reanalysis of these data (Steel and Wise 1979) indicates that controlling for semesters of mathematics taken virtually eliminates this male advantage. In the Fennema-Sherman studies (1977, 1978), which controlled for participation in mathematics courses, differences in favor of males were found in only two out of four high schools; in the middle school study, differences were found in favor of females in one of four school areas and in favor of males in another area.

This lack of opportunity to learn was also a factor in the first National Assessment of Educational Progress (NAEP), which gathered data in

1972–1973 from 9-, 13-, and 17-year-olds and from young adults. The male advantage on this first assessment has been described as overwhelming, particulary at older ages. At ages 9 and 13, when equal participation could be assumed, differences never exceeded six percentage points and some were in favor of girls (Mullis 1975). In the second NAEP assessment of performance in mathematics (NAEP 1979a, b), there was no significant difference between sexes on knowledge or skill items for 9- or 13-year-olds and only a difference of two percentage points on application items at age 13. In this later assessment, the average score for 17-year-old males was two to three percentage points higher on knowledge and skill items and five points higher on application items. Although statistically significant, these differences should be seen in the context of much larger differences correlated with race, region of the country, size and type of community, and parental education. While the situation may be improving and the ratio of males to females in higher-level secondary school mathematics courses may vary by locality, the NAEP reports indicate that overall the ratio is still three to two. Unlike the first assessment, the second did gather data on course participation, but the breakdown of performance differences at age 17 according to level of mathematics taken was not available when this chapter was written.

At least two other recent large-scale assessments have investigated sex-related performance differences of high school juniors and seniors with some degree of control for course participation. The 1978 California assessment (Law 1978) of students in grade twelve found that girls in all participation groups did better on computation, and boys in all groups did better on all types of measurement items except money items and on geometry applications, probability, and statistics. The pervasive difference on measurement items was also noted in the NAEP reports. The problem is that measurement skills are taught in scientific and vocational courses as well as in mathematics. The importance of considering this is evidenced by an assessment of high school juniors done in the state of Washington (DeWolf 1977). When students were grouped according to how much mathematics *and physics* they had taken, no sex-related differences in quantitative performance were found.

This rapid review of the recent history and present status of the issue of sex-related differences in mathematics achievement is also an index of the maturation of research methodology in the field. The broad generalizations of male superiority made a decade ago can be refined to these specifics:

1. The student population may have changed in ways that minimize sex-related differences.
2. Changes in tests due to raised consciousness about sex bias have eliminated some of the differences.
3. Most importantly, in studies that are controlled for number and type of mathematics courses taken, the size of the differences was small or nonexistent.

In summary, as of 1979 there seemed to be a few differences in *performance* in favor of males, which may be localized to certain schools or may be specific to mathematical tasks at higher cognitive levels. There are still substantial differences, however, in *participation* in mathematics courses. Possible sources of both of these types of differences will be explored.

RELATED FACTORS

While women have come a long way since the nineteenth century when biology was considered destiny and female proficiency in such things as higher mathematics was thought to lead to sterility or deformed offspring, biological sources of sex-related differences in mathematics achievement are still discussed in some quarters. The connection between sex and mathematics (especially problem solving) goes by way of another cognitive ability, called visual spatial ability or spatial visualization, and is measured by such tests as Space Relations of the Differential Aptitude Test Battery (Bennett, Seashore, and Wesman 1973). In the past, sex-related differences in spatial ability have appeared to parallel or precede those in mathematics (Maccoby and Jacklin 1974), and the use of charts, diagrams, and graphs in all branches of mathematics argues for the logic of this connection. A biological component of spatial ability was hypothesized because it seemed to have little to do with school training. Theories attributing differences in performance on spatial tests to genetic, metabolic, and hormonal differences between males and females have been described and evaluated by Sherman (1978) and generally found lacking. Furthermore, Sherman cites ample evidence of the effects of training on spatial ability. Although there is interesting work being done on hemispheric specialization and its relationship to spatial and mathematical tasks (Sherman 1978; Wheatley et al. 1978), differences in lateralization may have sociocultural or biological origins.

There is another fault with the argument that biology determines spatial ability, which in turn determines the ability to solve mathematical problems. If males were using their supposedly superior spatial skills to excel at solving problems, one would expect the correlations between the two types of abilities to be closer for males than for females. This is not the case. In some studies, such as those by Fennema and Sherman (1977, 1978) and others I reviewed (1979), there appeared to be no sex-related difference in the *relationship* between spatial and mathematical skills, even in the few groups in which there were sex-related differences in performance in either spatial or mathematical skills. In other studies, including mine (1979) and one by Sherman (1979), the relationship between spatial and mathematical variables appeared to be closer for *females* than for males. Given these difficulties with biological hypotheses, it seems that educators looking for sources of whatever sex-related differences in performance remain in 1979 should look to sociocultural factors instead.

Of the possible sociocultural factors influencing students' performance and participation in mathematics, some of the most important appear to be the expectations of significant others—parents, teachers, counselors, and peers. Research in this area has been summarized by Fox (1977). The review of studies of counselors cites examples of active discouragement of girls wishing to take advanced mathematics courses. Reasons given were that girls would do poorly and ruin their grade averages, careers in science and mathematics were too demanding for women, and science jobs were scarce and should go to men. There is some evidence that counselors' attitudes about women's roles are changing (Engelhard, Jones, and Stiggens 1976), but individual counselors should examine their own thinking for sex stereotyping in mathematics and science.

The review by Fox indicates that counselors were probably the least important "significant others" influencing participation in mathematics. This was supported by Kahl and Armstrong (1979), who did not find their scale, Counselor Encouragement to Take More Mathematics, to be among the highest correlates of actual amount of mathematics course participation among twelfth-grade students. What they did find as high correlates were students' perceptions of their mothers' and fathers' educational expectations for them and, to a lesser extent, their parents' encouragement. Fennema and Sherman (1977) also found significant correlations between mathematics achievement in grades nine through eleven and the student perceptions of their parents' attitudes toward them as learners of mathematics. Neither of these recent studies indicated that students perceived their fathers' attitudes as more influential than their mothers', as several studies reviewed by Fox (1977) suggest. At the middle school or junior high school level, student perceptions of parental attitudes were influential on both intended election of mathematics courses in the future (Kahl and Armstrong 1979) and on current performance (Fennema and Sherman 1978), although the correlation coefficients were smaller than in the high school studies. Furthermore, boys perceived their parents (especially their fathers) as supportive of their mathematical endeavors more often than girls did (Fennema and Sherman 1978).

Another "significant other" whose perceived influence was measured in both the Fennema-Sherman studies and the Kahl-Armstrong study was the mathematics teacher. In the latter study, student perception of teacher encouragement was influential in mathematics course participation for seniors and eighth graders, but perceptions of teachers' differential treatment of males and females or sex stereotyping were not. This variation from Fox's (1977) review of earlier research could be interpreted two ways: (a) teachers' consciousness has been raised on sex-related issues, or (b) students are ignoring teacher bias in this area. Fennema and Sherman (1978) reported significantly greater perception of teacher support by males than by females only in grade ten, but tenth grade is a crucial year for making decisions about further participation in mathematics.

Another possibility is that some teachers are behaving differently from others, but pooling the data, as was done in these large-scale studies, has obscured the differences. Parsons et al. (1979) found this to be true in their study of the effects of teacher expectancies on interaction in fifteen eighth- and ninth-grade mathematics classrooms. When they analyzed data from the five classrooms in which teachers had the most different expectations for boys and girls, they observed significantly more praise of boys' work than that of girls, a more public teaching style with fewer private student-teacher interactions, more criticism in general, and more reliance on volunteers for answers than in classes with little difference. In contrast, in the five classrooms with no sex-related differences in expectancies, girls interacted more and received more praise, and there was more one-to-one teacher-student interaction.

The importance of the mathematics teacher or individual classroom situation is underscored by the sequential nature of the subject. Anecdotal reports from mathematics autobiographies often project this idea: one bad experience and it is all over for mathematics. Casserly (1978), who has studied schools with twice the national ratio of girls to boys in advanced placement courses in mathematics, physics, and chemistry, reports that teachers in these schools are supportive but firm, refusing to excuse girls from participation in mathematics at the first sign of frustration or failure.

Last there is the influence of peers. Although Kahl and Armstrong (1979) did not find peer support important in actual or intended course participation for either sex, this conclusion contradicts the research reviewed by Fox (1977) and her own work with mathematically precocious females. She cites evidence, including some from Casserly, that special programs and advanced courses are most beneficial to females if the number of females does not become too small, although the "critical mass" is not known. One of the affective variables measured by Fennema and Sherman (1978) was student perception of mathematics as a male domain. Interestingly males stereotyped mathematics as a male domain significantly more than females at all grade levels, six through eleven, and in all areas of the city in those schools in which the study was carried out. One can only assume that the girls in those schools felt the weight of this male opinion. I majored in mathematics at a women's college in the early 1960s and vividly remember the response of a blind date who casually asked my major: "Mathematics? But you're a girl!"

So far the discussion of sociocultural sources of differential performance and participation has centered on external influences on the learner. Other factors to be considered are confidence (or anxiety) about mathematics and perceived usefulness of mathematics, which are internal influences although culturally determined. The Kahl-Armstrong (1979) research indicated the importance of confidence for both males and females in actual course participation by seniors and intended participation by younger students. Fennema and Sherman (1978) found confidence to be moderately related to mathematics achievement; there were sex-related differences in favor of males on this scale in grades eight through eleven, but they seemed to be localized to some schools or areas of the city.

The only other important factors in course participation identified by Kahl and Armstrong besides mathematics achievement centered around educational or career aspirations. This is to say that students taking mathematics perceived it to be useful in the future. Perceived usefulness of mathematics was significantly related to achievement for both males and females in grades six through eleven in the Fennema-

Sherman studies (1977, 1978), and the relationship was even closer for the small group of girls electing mathematics in grade twelve. Fox's (1977) review indicates that females are less interested in careers involving mathematics than are males, even females gifted in mathematics. Partly this seems to be a result of student perception that such careers are demanding and possibly incompatible with family responsibilities. Fox also reported positive results for girls from programs designed to present applications of mathematics to art and social issues. There seems to be a Catch 22 operating here. Females are shut out of certain careers and intellectual activities because they lack the necessary mathematical skills; they lack the skills because they did not continue to study mathematics, in part because they did not perceive mathematics as useful or their parents, teachers, or counselors did not perceive it as useful for them.

POSSIBILITIES FOR CHANGE

The school is the level at which to direct efforts toward change. Fennema and Sherman (1977) found sex-related differences in mathematics achievement in only two of the four high schools they studied, and Minuchin (1971) related school climate factors to sex-typed reactions and interactions of students. What then can teachers, counselors, and school administrators do to provide female students with an environment supportive of increased performance and participation in mathematics? Components of such an environment are indicated by the influences on participation and achievement discussed on the preceding pages.

An immediate concern should be eliminating or counteracting sex bias in textbooks, tests, and other instructional materials. Recently published articles indicate that some textbooks still portray mathematics as a predominantly male domain, and rating schemes available to evaluate textbooks and standardized tests being considered for adoption vary from the simple (Kepner and Koehn 1977) to the complex (Donlon, Ekstrom, and Lockheed 1979). Several strategies have been proposed for eliminating sex bias in instructional materials (Nibbelink and Mangru 1978). Teachers constructing tests could eliminate all references to people (often undesirable). They could use gender-free labels such as student or librarian, thus avoiding the use of the pronouns "he" and "she" (a strategy advocated by some publishers). Or they could use equal numbers of males and females randomly assigned to roles. Nibbelink and Mangru preferred the last because they found that children assigned a sex to the gender-free labels in a stereotyped way. The sex bias in mathematics materials that a teacher may have to use for other reasons can be a topic for class discussion. Although histories, such as Bell's *Men of Mathematics*, have been male dominated, there are now available books that present the contribution of women to mathematics (Osen 1974, Perl 1978).

Given the influence of significant others on females' participation and performance in mathematics, another priority is to reeducate counselors, teachers, and parents about its importance. Films, videotapes, and accompanying printed materials have been designed to help with this reeducation under the Women's Education Equity Act Program (WEEAP) and are being disseminated by the Educational Development Center. The Equals Project (Liff 1978) carried out workshops and developed materials for teachers, counselors, and administrators that stressed the usefulness of mathematics in a variety of careers and the importance of teaching problem solving in the mathematics classroom. Programs or classes are available that employ varying amounts of information, recreation, and therapy to help adults overcome negative feelings about mathematics (Afflack 1978; Liff 1978; Tobias 1978).

Removing negative pressure by peers from girls wanting to learn mathematics may necessitate intervention in the elementary school, where stereotyping mathematics as a male domain sometimes begins (Boswell 1979, Fox 1977). Casserly (1978) recommends early tracking so that mathematics-capable female students are not isolated. Single-sex classes have been advocated to relieve girls of the pressure of competing against boys, but separate has meant not equal for so long that this alternative is probably to be avoided. In upper-level courses, however, where the male-female ratio may be many-to-one, all the girls might be scheduled into one section rather than sprinkled over two or three. Casserly also recommends flexible programming so that students who drop out of mathematics in one grade can reenter, perhaps taking two courses in one year to catch up.

Mathematics teachers might change their modes of classroom instruction to facilitate cooperative, rather than competitive, activities. The research cited by Parsons et al. (1979) indicates that directing discussion by calling on students rather than waiting for volunteers and replacing some of the whole-group instruction with one-to-one interaction may also remove some of the peer pressure preventing females from actively pursuing mathematics. Given the fact the sex-related differences still found are on higher-level tasks requiring nonroutine responses, techniques such as brainstorming might be used to encourage risk-taking behavior. Teachers will need to stand ready, of course, to discourage stereotypic behavior even in these instructional modes; Hall (1976) noted that sexually mixed problem-solving teams appeared to lose input from female members because males dominated the conversation or a female was assigned to be the recorder or secretary. All these suggestions also apply to extracurricular activities in mathematics, which are often dominated by contests.

Teachers from other disciplines and counselors should discuss with students the amount of mathematics needed for careers in all areas: sciences, social sciences, humanities, business, and education. Emphasis on careers is especially needed for women whose mothers may not have been career oriented, because the social reality of the last quarter of the twentieth century is that women will work outside the home. Courses in social sciences, family life, and home economics should include discussion of the ways of combining career and family responsibilities. Adult female role models can be used as a source of inspiration and career information. Some of the WEEAP projects provide these on film and videotape; the Women and Mathematics Project of the Mathematics Association of America provides lists of women in science and mathematics who are willing to visit schools. Casserly (1978) also noted the effectiveness of having older female students in high school or college work with younger students who may admire an adult career woman but may not have much of an idea of how to get from their position to hers.

The current level of federal funding of projects on women and mathematics and the level of interest in the topic in professional associations indicate the distance traveled since a research report of a large, federally funded project contained the following remark on observed sex-related differences: "Interpretation and comment on this pattern will be left to persons involved in the women's liberation movement" (Wilson 1972, p. 95). In the past ten years researchers have concentrated on defining the problem and identifying its sources. Educators have only begun to propose solutions and test their effectiveness. It is clear that mathematics performance and participation are woven in a complex pattern into the fabric of society's expectations and its roles for women and men. The changes that have occurred, however, suggest that further change is possible to ensure equal participation in society's economic and intellectual activities, regardless of sex.

BIBLIOGRAPHICAL NOTE

For the person interested in pursuing this topic further than the limits of this chapter, I suggest two books in particular. The first, *Women and Mathematics: Research Perspectives for Change*, contains reviews written by Fox, Fennema, and Sherman of research published before 1977 as well as extensive bibliographies. Originally published by the National Institute for Education (NIE), it is now available only from ERIC. The second is *Perspectives on Women and Mathematics*, edited by J. E. Jacobs, which contains research summaries and bibliographies, descriptions of courses and projects, and lists of contacts and materials. It is available through ERIC or from the National Council of Teachers of Mathematics. Most research reports cited below as papers presented at the 1979 annual meetings of professional organizations are part of a group of studies funded by NIE. Abstracts of these proposals and authors' addresses are available from NIE, and final reports can be requested from the authors. The National Science Foundation has also funded studies on women and mathematics. Brochures describing the WEEAP materials are available from the Educational Development Center, 55 Chapel St., Newton, MA, 02160. Many state departments of education now do their own assessments of mathematical progress and may have data analyzed by sex.

REFERENCES

Afflack, Ruth. "A Mini-course for the Mathphobic." In *Perspectives on Women and Mathematics*, edited by Judith E. Jacobs. Columbus, Ohio: ERIC Information Analysis Center for Science, Mathematics, and Environmental Education, 1978.

Bennett, George K.; Seashore, Harold G.; and Wesman, Alexander C. *Differential Aptitude Tests: Administrator's Handbook.* New York: Psychological Corporation, 1973.

Boswell, Sally L. "Sex Roles, Attitudes, and Achievement in Mathematics: A Study of Elementary School Children and Ph.D.'s." Paper presented at a meeting of the Society for Research in Child Development, San Francisco, March 1979.

Carnegie Commission on Higher Education. *Opportunities for Women in Higher Education.* New York: McGraw-Hill Book Co., 1973

Casserly, Patricia L. "Factors Leading to Success: Present and Future." In *Perspectives on Women and Mathematics*, edited by Judith E. Jacobs (Columbus, Ohio: ERIC Information Analysis Center for Science, Mathematics, and Environmental Education, 1978.

DeWolf, Virginia A. *High School Mathematics Preparation and Sex Differences in Quantitive Abilities.* Seattle: Educational Assessment Center, University of Washington, 1977. ERIC: ED 149 968.

Donlon, Thomas F. *Content Factors in Sex Differences on Test Questions.* Princeton, N.J.: Educational Testing Service, 1973.

Donlon, Thomas F.; Ekstrom, Ruth B.; and Lockheed, Marlaine E. "The Consequences of Sex Bias in the Content of Major Achievement Test Batteries." *Measurement and Evaluation in Guidance* 11 (January 1979): 202–16.

Engelhard, Patricia A.; Jones, Kathryn O.; and Stiggens, Richard J. "Trends in Counselor Attitude about Women's Roles." *Journal of Counseling Psychology* 23 (July 1976): 365–72.

Fehr, Howard F. "The Secondary School Mathematics Curriculum Improvement Study: A Unified Mathematics Program." *Mathematics Teacher* 67 (January 1974): 25–33.

Fennema, Elizabeth. "Mathematics Learning and the Sexes: A Review." *Journal for Research in Mathematics Education* 5 (May 1974): 126–39.

Fennema, Elizabeth. "Sex-related Differences in Mathematics Achievement: Where and Why?" In *Perspectives on Women and Mathematics*, edited by Judith E. Jacobs. Columbus, Ohio: ERIC Information Analysis Center for Science, Mathematics, and Environmental Education, 1978.

Fennema, Elizabeth, and Sherman, Julia A. "Sex-related Differences in Mathematics Achievement, Spatial Visualization, and Affective Factors." *American Educational Research Journal* 14 (Winter 1977): 51–71.

Fennema, Elizabeth, and Sherman, Julia A. "Sex-related Differences in Mathematics Achievement and Related Factors: A Further Study." *Journal for Research in Mathematics Education* 9 (May 1978): 189–203.

Flanagan, John C. "Changes in School Levels of Achievement: Project TALENT Ten and Fifteen Year Retests." *Educational Researcher* 5 (September 1976): 9–12.

Flanagan, John C.; Davis, Frederick B.; Daily, John T.; Shaycroft, Marion F.; Orr, David B.; Golberg, Isadore; Neyman, Clinton A., Jr. *The American High School Student Today.* Pittsburgh: University of Pittsburgh, 1964.

Fox, Lynn H. "The Effects of Sex Role Socialization on Mathematics Participation and Achievement." In *Women and Mathematics: Research Perspectives for Change*, edited by Lynn H. Fox, Elizabeth Fennema, and Julia Sherman. NIE Papers in Education and Work, no. 8. Washington, D.C.: U.S. Department of Health, Education, and Welfare, 1977.

Hall, Thomas R. "A Study of Situational Problem Solving by Gifted High School Mathematics Students." Doct. diss., Georgia State University at Atlanta, 1976.

Kahl, Stuart R., and Armstrong, Jane. "Correlates of Mathematics Course Participation in High School." Paper presented at the Annual Meeting of the American Educational Research Association, San Francisco, April 1979.

Kepner, Henry S., Jr., and Koehn, Lilane R. "Sex Roles in Mathematics: A Study of the Status of Sex Stereotypes in Elementary Mathematics Texts." *Arithmetic Teacher* 24 (May 1977): 379–85.

Law, Alexander I. *Student Achievement in California Schools: 1977–1978, Annual Report (Mathematics Section).* Sacramento, Calif.: California State Department of Education, 1978.

Liff, Rita May. "Programs to Combat Math Avoidance." In *Perspectives on Women in Mathematics*, edited by Judith E. Jacobs. Columbus, Ohio: ERIC Information Analysis Center for Science, Mathematics, and Environmental Education, 1978.

Maccoby, Eleanor, and Jacklin, Carol N. *The Psychology of Sex Differences.* Stanford, Calif.: Stanford University Press, 1974.

Minuchin, Patricia P. "Sex-role Concepts and Sex Typing in Childhood as a Function of School and Home Environments." In *Social Development and Personality*, edited by George G. Thompson. New York: Wiley, 1971.

Mullis, Ina V. S. *Educational Achievement and Sex Discrimination.* Denver, Colo.: National Assessment of Educational Progress, 1975.

National Assessment of Educational Progress. *Mathematical Knowledge and Skills.* NAEP Report No. 09-MA-02. Washington, D.C. U.S. Government Printing Office, 1979(a).

National Assessment of Educational Progress. *Mathematical Applications.* NAEP Report No. 09-MA-03. Washington, D.C.: U.S. Government Printing Office, 1979(b).

Nibbelink, William H., and Mangru, Matadial. "Sexism in Mathematics Textbooks." In *Perspectives on Women in Mathematics*, edited by Judith E. Jacobs. Columbus, Ohio: ERIC Information Analysis Center for Science, Mathematics, and Environmental Education, 1978.

Osen, Lynn. *Women in Mathematics.* Cambridge, Mass.: Masachusetts Institute of Technology, 1974.

Parsons, Jacquelynne E.; Heller, Kirby A.; Meece, Judith L.; Kaczola, Carol. "The Effects of Teachers' Expectancies and Attributions on Students' Expectancies for Success in Mathematics." Paper presented at the Annual Meeting of the American Educational Research Association, San Francisco, April 1979.

Perl, Teri H. *Math Equals.* Reading, Mass.: Addison-Wesley, 1978).

Schonberger, Ann K. "Are Mathematics Problems a Problem for Women and Girls?" In *Perspectives on Women in Mathematics*, edited by Judith E. Jacobs. Columbus, Ohio: ERIC Information Analysis Center for Science, Mathematics, and Environmental Education, 1978.

Schonberger, Ann K. "The Relationship between Visual Spatial Abilities and Mathematical Problem Solving: Are There Sex-Related Differences?" In *Proceedings of the Third International Conference for the Psychology of Mathematics Education*, edited by David O. Toll. Warwick, England: The Warwick Mathematics Education Research Centre, 1979.

Sherman, Julia A. *Sex-related Cognitive Differences: An Essay on Theory and Evidence.* Springfield, Ill.: Charles C. Thomas, 1978.

Sherman, Julia A. "Predicting Mathematics Performance in High School Girls and Boys." *Journal of Educational Psychology* 71 (April 1979): 242–49.

Steel, Lauri, and Wise, Lauress L. "Origins of Sex Differences in High School Mathematics Achievement and Participation." Paper presented at the Annual Meeting of the American Educational Research Association, San Francisco, April 1979.

Tobias, Sheila. *Overcoming Math Anxiety.* New York: W. W. Norton and Co., 1978.

Wheatley, Grayson H.; Mitchell, Robert; Frankland, Robert L.; and Kraft, Rosemarie. "Hemispheric Specialization and Cognitive Development: Implications for Mathematics Education." *Journal for Research in Mathematics Education* 9 (January 1978): 20–32.

Wilson, James W. "Patterns of Achievement in Grade 11: Z Population." *NLSMA Reports*, no. 17, edited by James W. Wilson and Edward Begle. Stanford, Calif.: School Mathematics Study Group, Stanford University Press, 1972.

12. The Teacher Variable in Mathematics Instruction

Douglas A. Grouws

There is now strong research support for the fact that teachers make a difference in the amount of learning that takes place in classroom situations (Good, Biddle, and Brophy 1975; Rakow, Airasian and Madaus 1978). It is also quite clear that some teachers are much more effective than others in their impact on student achievement. Researchers, for example, have consistently been able to distinguish highly effective teachers from less effective teachers (Brophy and Evertson 1974; Berliner and Tikunoff 1976; McDonald and Elias 1976). In a study of fourth-grade mathematics instruction, Good and Grouws (1975a, 1977) identified nine effective and nine less-effective teachers from a sample of forty-one teachers. Over a three-year period, the effective teachers consistently produced better than expected results in mathematics achievement (residualized gain scores), while the less-effective teachers consistently produced less than predicted achievement gains. Classroom observation of these eighteen teachers revealed different patterns of behavior for each of the two groups of teachers. The particular clusters of behavior associated with the effective group are discussed in the section of this chapter on teacher behavior. First, however, the important consideration of how stable a teacher's impact on students is from year to year is discussed.

STABILITY OF TEACHER PERFORMANCE

It is worthwhile to note that while some teachers are more effective than others, teacher impact from year to year tends to vary considerably even when the initial achievement differences from one year's class to the next year's class are taken into account (Brophy 1973; Emmer, Evertson, and Brophy 1979). In mathematics, for example, Good and Grouws found overall teacher stability across a three-year period to be quite low, although teachers at the extremes of the distribution seemed to be much more stable than those in the middle of the distribution. This lack of stability also seems to hold true in mathematics contexts when attitude is used as the criterion (Good and Grouws 1975b). The reason for this lack of impact stability is not known, but Good (1979) suggests that one explanation for the lack of consistency may be that the relative effectiveness of teachers with minimal skills in a given year may depend upon subtle context variables or on circumstances in the personal lives of teachers that alter the amount of time that they can devote to instructional planning and preparation. An important first step in studying stability of teacher impact would be to determine whether the *instructional behavior* of teachers tends to be stable over time and contexts.

ACCOUNTING FOR DIFFERENCES IN TEACHER IMPACT

Given that teachers make a difference in the mathematics achievement of students, what accounts for some teachers being more effective than others? Is it the teacher's characteristics, the teacher's feelings, or what the teacher does? Characteristics of teachers, such as age, amount of teaching experience, educational background, professional activity, and so on, have been extensively studied in the past. Based on this research, it seems reasonable to conclude that teacher characteristics have very little relationship to effective instruction in mathematics. This conclusion has also been drawn in several reviews of mathematics teaching (Fey 1969; Begle 1979; Cooney 1980); hence, it is a bit surprising to find many people who still look to this area when considering teaching effectiveness. It is definitely time to look beyond these status variables in the search for the variables associated with teacher effectiveness.

An area that has shown considerable promise in recent years is the study of teacher behaviors. Many of the behaviors that have been identified as being associated with effective teaching have come from what are commonly called process-product studies. In this type of research those teacher behavior variables that seem to hold potential are identified and operationally defined. The frequency and extent of their occurrence is then determined in many classrooms over a fixed period of time. Finally, the correlation between the measures of the variables and the average student achievement scores (residualized gain scores) during the observation period are computed. A statistically significant positive correlation between one of the variables and mathematics achievement suggests that effective teachers use this variable more often. Replication of the result in subsequent studies gives credibility to the finding and leads to the examination of the variable in field-based experimental studies where cause-and-effect relationships can be determined. In the last few years, a number of experimental studies based on process-product findings in mathematics have been designed and conducted.

IMPORTANT TEACHER BEHAVIORS

Several studies of the process-product type have focused on the teaching of mathematics. In the previously mentioned study of fourth-grade mathematics, Good and Grouws (1977) found teacher effectiveness to be strongly associated with the following behavioral clusters: (a) general clarity of instruction, (b) task-focused environment, (c) nonevaluative (comparatively little use of praise and criticism) and relatively relaxed learning environment, (d) higher achievement expectations (more homework, faster pace), (e) relatively few behavioral problems, and (f) the class taught as a unit. In a carefully selected, small sample of seventh- and eighth-grade mathematics teachers, Evertson, Emmer, and Brophy (1980) found that—in contrast to less effective teachers—more effective teachers (a) spent more time on content presentations and discussions and less time on individual seatwork; (b) manifested behaviors indicative of higher expectations of their students (assigned homework more frequently, stated concern for academic achievement, and gave academic encouragement more often); and (c) exhibited stronger management skills (minimized inappropriate behavior, made more efficient transitions, and received more student attentiveness). In a study of instruction in algebra, Smith (1977) found that more effective teachers used fewer vague terms (for example, chances are, sometimes, almost, often), used more relevant examples, and gave greater attention to lesson objectives. The study by Smith is particularly noteworthy in that clarity, a variable that has been identified in many general studies (Rosenshine and Furst 1973), was operationally defined in a mathematics setting in such a way that the associated behaviors were easily observed and quantified.

In a number of studies in subject matter areas other than mathematics, other variables such as enthusiasm, warmth, and variability have also been associated with effective teaching. There is a significant need for such general variables to be operationally defined in a mathematics setting and then examined in an experimental framework to determine if there is a causal relationship between the variables and mathematics learning gains. Cooney (1980) points out that it is largely unknown whether these variables are general in nature or subject-matter specific. It may well be that affective variables such as warmth and enthusiasm cut across subject matter areas, while cognitive variables such as clarity and variability have distinct or special meanings in mathematics that would not hold true in other subjects, and thus these variables may have a more pronounced effect in mathematics than in other disciplines. The possibility of certain behaviors having special effects on particular subject matter areas has some research support from a study by Evertson, Anderson, and Brophy (1978). They found that behaviors highly correlated with teacher effectiveness in mathematics did not correlate with teacher effectiveness in English. They attributed this primarily to a difference in the types of instructional goal for these two subject-matter areas.

INSTRUCTIONAL TIME

The use of instructional time in conceptualizing school learning is grounded in the work of Carroll (1963) and the more recent work of Wiley and Harnischfeger (1974) and Bloom (1976). The Beginning Teacher Evaluation Study (BTES) (McDonald 1976; Fisher et al. 1978) and a number of other studies taken together lend strong support to several generalizations. First, the more time provided for mathematics instruction, the more mathematics the students learn. Thus the number of school days, the number of mathematics periods per week, the length of the mathematics period, and the amount of time spent on mathematics during a class period significantly affect the amount of mathematics learned. In view of this finding, it is most unfortunate that in some school situations mathematics is not taught on a daily basis or is taught for a relatively short period of time each day. For example, in the BTES study the average amount of time spent in second-grade mathematics instruction varied from twenty-five minutes per day in one classroom to over sixty minutes per day in another classroom. The implications of such time allocation decisions for students' achievement in mathematics are very clear.

The second generalization that can be drawn is that the amount of the students' *engaged time* is positively related to the amount of learning. That is, the amount of time that students are on-task (processing information, listening, manipulating, reading, thinking) during the time allocated to mathematics is directly related to their gains in mathematical learning. This conclusion is not too surprising, but the variance in the amount of engaged time found in mathematics classrooms is quite surprising. For example, in the BTES study some mathematics classes averaged a 50 percent engagement rate while other classes averaged over a 90 percent engagement rate. Clearly, teachers should be cognizant of engaged time as they consider their classroom instruction.

The final generalization is that students' success rate on classroom tasks is related to achievement. Fisher, Marliave, and Filby (1979) found that mathematics students in grade five who spent more time than the average in high-success activities (error rates of less than 20 percent) also had higher year-end achievement scores, better retention of learning over the summer, and more positive attitudes toward school. They defined time spent on tasks with error rates less than 20 percent as academic learning time (ALT). The effect of increases in ALT on certain measures of mathematics achievement is not denied; however, there is a very urgent need to measure the effect of such increases on various aspects of mathematics achievement, including concepts, skills, and problem solving as well as students' attitudes toward mathematics. Given the emphasis of many standardized tests on skill-oriented objectives, it may be that the gains produced by increased ALT are in the skill areas and may even come at the expense of improvement in the problem-solving area. This is particularly true given the generally accepted maxim that students must frequently be engaged in problem-solving activity to be good problem solvers. In making this statement most mathematics educators are thinking of problems as situations that are new to the student and where the path to the solution is not immediately obvious. It seems clear, then, that tasks of this type would frequently involve error rates of greater than 20 percent.

A time-related variable that is worthy of additional study is "wait time." Wait time is the amount of time a teacher waits after calling on a student for the student to respond. Rowe (1978) found in science classes the wait times were often extremely short and that increasing wait times tended to result in increases in appropriate responses and students' confidence and decreases in failures to respond and managerial problems.

DIRECT INSTRUCTION

Attempts to summarize teaching effectiveness research by looking beyond the isolating of specific behaviors to clusters of related behaviors that have consistent predictable effects have recently been quite productive. A pattern of behaviors common to effective teaching across many studies has been identified by a number of reviews of research (Good 1979; Gage 1978; Medley 1977; Rosenshine 1979). This pattern of behaviors, most often referred to as direct instruction, is characterized by Rosenshine (1979) as involving an academic focus, little student choice of activity, use of large-group instruction, and use of factual questions and controlled practice in instruction. Good (1979) describes direct instruction as when "a teacher sets and articulates the learning goals, actively assesses student progress, and frequently makes class presentations illustrating how to do assigned work." Although these descriptions and others are all somewhat different, the concept of direct instruction does provide a useful framework for examining past research and also a useful guideline for practicing teachers as they plan instruction and examine their own teaching behavior.

Three experimental studies (Anderson, Evertson, and Brophy 1979; Good and Grouws 1979a, 1979b) have assessed instructional treatments that seem to have much in common with the concept of direct instruction. The Anderson, Evertson, and Brophy (1979) study presented twenty-two specific principles to first-grade teachers for implementation in their reading instruction. Most parts of the treatment were implemented, and at the end of the experimental period, the students in the treatment condition had significantly higher scores on achievement than did the students in the control condition where teachers taught in their usual way. Another interesting result of this study was that there were no differences between the observed and the unobserved treatment groups. This suggests that in-class observation is not necesary to insure treatment implementation, thus greatly increasing the external validity of the study.

One of the Good and Grouws (1979a) studies involved forty fourth-grade mathematics classes in the experimental testing of a system of instruction that had many aspects similar to those of direct instruction. Their system of instruction involved the following: (a) instructional activity was initiated and reviewed in the context of meaning; (b) a substantial portion of each lesson was devoted to content development (the focus was on the teacher actively developing ideas, conveying meanings, giving examples, and so on); (c) students were prepared for each lesson stage to enhance involvement; and (d) the principles of distributed and successful practice were employed. Pre- and posttesting with a standardized achievement test indicated that after two and one-half months of the treatment the performance of students in the experimental group was substantially better than that of those in control classrooms. End-of-year achievement testing by the school district indicated that experimental classes continued to perform better than control classes three months after the posttesting on the mathematics subtests of a standardized achievement test. Also, experimental students had significantly better attitudes toward mathematics than did control students at the end of the treatment period, as measured on a ten-item attitude scale. Thus, the concept of direct instruction seems to have both cognitive and affective potential in mathematics classrooms. Of course, additional research using this concept in mathematics settings is needed to answer a plethora of questions.

Peterson (1979) has pointed out the need to examine the types of students that benefit most from direct instruction and to determine the kinds of learning outcomes best fostered by direct instruction. There has been some work on direct instruction in mathematics that relates to the former point. Good and Grouws (1979a, 1979b) found that direct instruction in their study did *not* have a significant differential impact across students at different levels of achievement. In a further analysis of these data, Ebmeier and Good (1979) found a trend within direct instruction toward benefiting certain types of students more than others when student types were defined both in terms of achievement and personality variables. In particular, Ebmeier and Good's data suggest that high-ability/task-oriented students may benefit more from direct instruction than do students of other types. Further research is definitely needed to determine the practical impact of direct instruction on different types of students and to ascertain ways in which a direct instruction model might be modified to accommodate more fully particular types of students.

Mathematics achievement can appropriately be considered as the composite of a number of different outcomes including concepts, skills, and problem solving. Thus, it is important that the effects of direct instruction be assessed across a broad spectrum of these outcomes. To date, research data support direct instruction as having a strong impact on general mathematics achievement. In their fourth-grade study, Good and Grouws (1979a) note that gains from direct instruction were greatest in the skill and concept areas.

A subsequent study (Good and Grouws, 1979b) assessed the impact of direct instruction in sixth-grade mathematics classrooms where the direct-instruction model was adjusted to give special attention to verbal problem solving through the daily use of special techniques such as estimating answers, writing verbal problems, and so on. The results of the study showed that problem-solving performance in the treatment group was significantly better than in the control group. The gains, however, seemed to come at the expense of progress in the concept and skill areas. Hence, it appears that the direct-instruction model can be adapted to generate differential impact on a variety of educational outcomes, including basic skills and problem solving. A study by Weber (1978) in ninth-grade algebra classes tends to support this conclusion. Her research indicates that when the teacher actively structures learning through lecture and feedback, high levels of achievement for both low and high mental processes are produced.

Further research on the effects of a direct-instruction model on a variety of outcomes is needed. This research could profitably include a focus on problem solving (defined to include a broad range of problems), creativity, and other high-level cognitive outcomes. Progress is definitely needed on ways to fine-tune this system of instruction to meet specific outcomes for particular kinds of students without generating an overload on the instructional demands placed on the teacher.

The potential for the concept of direct instruction is great, but already the usefulness of the concept is being impaired by loosely comparing and contrasting it with terms such as open teaching, traditional teaching, and indirect instruction that have come to have such broad meaning as to have little meaning at all. It is worthwhile to recall how the meaning of "discovery learning" became diluted in this way and to avoid letting this happen to direct instruction.

The potential of direct instruction as an orienting concept for teachers and a guidepost for researchers can best be realized if several generalizations are constantly considered. First, direct instruction encompasses the entire instructional period. It is not limited to the discourse portion of the lesson as are concepts such as indirect teaching and teacher-centered instruction, and, hence, it can be viewed best as a system of instruction. Second, direct instruction involves a substantial amount of active teaching with a focus on meaning and comprehension. In mathematics, this has become known as the development portion of the lesson and devoting over one-half of class time to it has strong research support across grade levels (Shuster and Pigge 1965; Shipp and Deer 1960; Zahn 1966; Dubriel 1977). This active teaching can involve a number of teaching methods, including lectures, discovery lessons, demonstrations, and so on.

Third, direct instruction requires attention to and active monitoring of student activities. In mathematics, this means careful consideration and supervision of practice activities, laboratory lessons, and so on. Fourth, direct instruction should be viewed *not* as a specific set of behaviors that must be adhered to but rather as an interactive system effective under a number of conditions upon several kinds of learning outcomes. But it is probable that direct instruction must be sensitively modified to meet the needs of particular types of students and to achieve specific learning outcomes. Effective use of direct instruction may reduce behavioral problems and influence other aspects of classroom management.

MANAGEMENT VARIABLES

The role of the teacher as classroom manager is a critical aspect of effective teaching. Good (1979) points out that "teachers' managerial abilities have been found to relate positively to student achievement in every process-product study to date [p. 54]." These managerial responsibilities cover a host of activities. Aside from the control of behavioral problems, some of the more important managerial decisions seem to revolve around teachers' academic expectations of their students. For example, the amount and type of homework required of students seems to be important (Good and Grouws 1977); the amount of subject matter material presented (Porter et al. 1978), and the instructional pace (rate) through the curriculum (Good, Grouws, and Beckerman 1978) seem to affect achievement. Many studies have examined these ideas and qualified them in order to heighten their impact on students' achievement. Importantly, each of these ideas seems to be directly related to the idea of "opportunity to learn." That is, each is linked to the idea that students must be exposed (sometimes repeatedly) to an idea or a skill before it can be learned. However, it seems that in many instances when an idea is taught, it is frequently learned to a reasonable extent by most students. Clearly this does not suggest that all teachers should immediately begin to increase the rate at which they move through the curriculum. It does suggest, however, that many teachers could profitably consider their pace in connection with their particular students and situation in order to enhance their effectiveness.

SUMMARY

Teachers play an important and complex role in mathematics instruction. They have a significant impact on how much students learn and on what they learn. Some teachers are consistently more effective than others, although the extent of a given teacher's influence on students' learning may vary from year to year.

Recent research provides strong evidence to support the effectiveness of certain teaching practices or clusters of behavior; yet as might be expected, no single behavior in isolation has been causally related to improved performance for all students under all conditions. Increasing instructional time devoted to mathematics and increasing the amount of time students are actively attending to instruction or to instruction-related tasks is a research-supported guideline for increasing achievement in mathematics. Another guidepost for effective teaching is direct instruction. Teachers should consider the importance of actively teaching and structuring ideas, carefully monitoring students' activities and practice, and providing for distributed and successful practice and the other aspects of a direct-instruction model. This consideration must take into account the types of students the teacher has and the kinds of learning outcomes desired. Either or both of these situations may indicate the need for and direction of appropriate modifications in the direct instruction model. Finally, to increase classroom effectiveness a teacher should be influenced by the opportunity to learn variable and related behaviors, such as pacing, that seem to influence student achievement.

Fortunately, our understanding of mathematics education has advanced to the point that some data-supported generalizations could be recommended for consideration. This is an important step forward, yet much research and critical thinking needs to be done so that suggestions and implications can be more specific and therefore more helpful.

REFERENCES

Anderson, Linda; Evertson, Carolyn; and Brophy, Jere E. "An Experimental Study of Effective Teaching in First-grade Reading Groups." *Elementary School Journal* 79 (March 1979): 193–223.

Begle, Edward G. *Critical Variables in Mathematics Education.* Washington, D.C.: Mathematics Association of America and the NCTM, 1979.

Berliner, David, and Tikunoff, William J. "The California Beginning Teacher Evaluation Study: Overview of the Ethnographic Study." *Journal of Teacher Education* 27 (Spring 1976): 24–30.

Bloom, Benjamin S. *Human Characteristics and School Learning.* New York: McGraw-Hill, 1976.

Brophy, Jere. "Stability of Teacher Effectiveness." *American Educational Research Journal* 10 (Summer 1973): 245–52.

Brophy, Jere, and Evertson, Carolyn. *Process-product Correlations in the Texas Teacher Effectiveness Study: Final Report.* Report no. 74-4. Austin, Tex: Research and Development Center for Teacher Education, University of Texas, 1974.

Carroll, John B. "A Model of School Learning." *Teachers College Record* 64 (May 1963): 723–33.

Cooney, Thomas J. "Research on Teaching and Teacher Education." In *Research in Mathematics Education*, edited by Richard Shumway. NCTM Professional Reference Series. Reston, Va.: NCTM, 1980.

Dubriel, John B. "A Study of Two Plans for Utilization of Class Time in First-Year Algebra." Doct. diss., University of Missouri, Columbia, 1977.

Ebmeier, Howard, and Good, Thomas L. "The Effects of Instructing Teachers about Good Teaching on the Mathematics Achievement of Fourth Grade Students." *American Educational Research Journal* 16 (Winter 1979): 1–16.

Emmer, Edmund T.; Evertson, Carolyn; and Brophy, Jere E. "Stability of Teacher Effects in Junior High Classrooms." *American Educational Research Journal* 16 (Winter 1979): 71–75.

Evertson, Carolyn; Anderson, Linda; and Brophy, Jere E. *Texas Junior High School Study: Final Report of Process-Outcome Relationships.* Vol. 1. Report no. 4601. Austin, Tex.: Research and Development Center for Teacher Education, University of Texas, 1978.

Evertson, Carolyn; Emmer, Edmund T.; and Brophy, Jere E. "Predictors of Effective Teaching in Junior High Mathematics Classrooms." *Journal for Research in Mathematics Education* 11 (May 1980): 167–78.

Fey, James. "Classroom Teaching of Mathematics." *Review of Educational Research* 39 (October 1969): 535–51.

Fisher, Charles W.; Filby, Nikola; Marliave, Richard N.; Cahen, Leonard; Dishaw, Marilyn; Moore, Jeffrey; and Berliner, David C. *Teaching Behaviors, Academic Learning Time and Student Achievement: Final Report of the Phase III-B Beginning Teacher Evaluation Study.* San Francisco: Far West Laboratory for Educational Research and Development, 1978.

Fisher, Charles W.; Marliave, Richard N.; and Filby, Nikola N. "Improving Teaching by Increasing 'Academic Learning Time'." *Educational Leadership* 37 (October 1979): 52–54.

Gage, N. L. *The Scientific Basis of the Art of Teaching.* New York: Teachers College Press, Columbia University, 1978.

Good, Thomas L. "Teacher Effectiveness in the Elementary School." *Journal of Teacher Education* 30 (March-April 1979): 52–64.

Good, Thomas L.; Biddle, Bruce; and Brophy, Jere E. *Teachers Make a Difference.* New York: Holt, Rinehart & Winston, 1975.

Good, Thomas L., and Grouws, Douglas A. *Process-Product Relationships in Fourth-Grade Mathematics Classrooms.* Final Report of National Institute of Education Grant NE-G-00-3-0123. Columbia: University of Missouri, December 1975(a).

Good, Thomas L., and Grouws, Douglas A. "Teacher Rapport: Some Stability Data." *Journal of Educational Psychology* 67 (April 1975): 179–82(b).

Good, Thomas L., and Grouws, Douglas A. "Teaching Effects: A Process-Product Study in Fourth-Grade Mathematics Classrooms." *Journal of Teacher Education* 28 (May-June 1977): 49–54.

Good, Thomas L., and Grouws, Douglas A. "The Missouri Mathematics Effectiveness Project: An Experimental Study in Fourth-Grade Classrooms." *Journal of Educational Psychology* 71 (June 1979): 355–62(a).

Good, Thomas L., and Grouws, Douglas A. *Experimental Study of Mathematics Instruction in Elementary Schools.* Final Report of National Institute of Education Grant NIE-G-77-003. Columbia: University of Missouri, December 1979(b).

Good, Thomas L.; Grouws, Douglas A.; and Beckerman, Terrill. "Curriculum Pacing: Some Empirical Data in Mathematics." *Journal of Curriculum Studies* 10 (January-March 1978): 75–81.

McDonald, Frederick J. "Report on Phase II of the Beginning Teacher Evaluation Study." *Journal of Teacher Education* 27 (Spring 1976): 39–42.

McDonald, Frederick J., and Elias, Patricia. *The Effects of Teacher Performance on Pupil Learning.* Beginning Teacher Evaluation Study, Phase II, Final Report. Vol. 1. Princeton, N.J.: Educational Testing Service, 1976.

Medley, Donald M. *Teacher Competence and Teacher Effectiveness: A Review of Process-Product Research.* Washington, D.C.: American Association of Colleges for Teacher Education, 1977.

Peterson, Penelope L. "Direct Instruction: Effective for What and for Whom?" *Educational Leadership* 37 (October 1979): 46–48.

Porter, Andrew C.; Schmidt, William; Floden, Robert; and Freeman, Donald. *Impact on What: The Importance of Content Covered.* Research series no. 2. East Lansing, Mich.: Institute for Research in Teaching, Michigan State University, 1978.

Rakow, Ernest A.; Airasian, Peter; and Madaus, George F. "Assessing School and Program Effectiveness: Estimating Teacher Level Effects." *Journal of Educational Measurement* 15 (Spring 1978): 15–21.

Rosenshine, Barak. "Content, Time, and Direct Instruction." In *Research on Teaching: Concepts, Findings, and Implications,* edited by Penelope L. Peterson and Herbert J. Walberg. Berkeley, Calif.: McCutchan Publishing Corp., 1979, pp. 28–56.

Rosenshine, Barak, and Furst, Norma. "The Use of Direct Observation to Study Teaching." In *Second Handbook of Research on Teaching*, edited by Robert M. W. Travers. Chicago: Rand McNally, 1973, pp. 122–83.

Rowe, Mary B. "Wait, Wait, Wait. . . ." *School Science and Mathematics* 78 (March 1978): 207–16.

Shipp, Donald, and Deer, George H. "The Use of Class Time in Arithmetic." *Arithmetic Teacher* 7 (March 1960): 117–21.

Shuster, Albert H., and Pigge, Fred L. "Retention Efficiency of Meaningful Teaching." *Arithmetic Teacher* 12 (January 1965): 24–31.

Smith, Lyle R. "Aspects of Teacher Discourse and Student Achievement in Mathematics." *Journal for Research in Mathematics Education* 8 (May 1977): 195–204.

Weber, Margaret B. "The Effect of Learning Environment on Learner Involvement and Achievement." *Journal of Teacher Education* 29 (November-December 1978): 81–85.

Wiley, David E., and Harnischfeger, Annegret. "Explosion of a Myth: Quantity of School and Exposure to Instruction, Major Educational Vehicles." *Educational Researcher* 3 (April 1974): 7–12.

Zahn, Karl G. "Use of Class Time in Eighth Grade Arithmetic." *Arithmetic Teacher* 13 (February 1966): 113–20.

PART FIVE
Today and Tomorrow

13. National Assessment: A Perspective of Students' Mastery of Basic Mathematics Skills

Thomas P. Carpenter, Mary Kay Corbitt, Henry S. Kepner, Jr., Mary Montgomery Lindquist, and *Robert E. Reys*

In recent years a great deal of attention has been focused on the teaching of basic skills. In part because of reports of declining scores on standardized tests (Advisory Panel on the Scholastic Aptitude Test Score Decline 1977), there seems to be a general feeling that many students are inadequately prepared in basic skills in mathematics and other content areas. Although such reports frequently paint a bleak picture of students' overall performance, they seldom provide any explicit insights as to what skills students are or are not learning. Perhaps the best source of specific information regarding students' knowledge of basic skills in mathematics is the mathematics assessment of the National Assessment of Educational Progress (NAEP).

National Assessment conducted its second mathematics assessment during the 1977–78 school year. Exercises covering a wide range

This chapter is based upon work supported by the National Science Foundation under Grant No. SED-7920086. Any opinions, findings, and conclusions or recommendations expressed in this publication are those of the authors and do not necessarily reflect the views of the National Science Foundation.

of objectives were administered to a carefully selected, representative sample consisting of more than seventy thousand 9-, 13-, and 17-year-olds. The results provide an accurate sampling of the knowledge of American elementary and secondary students rather than a special population such as college-bound seniors. The assessment also has the advantage of providing analysis of performance on specific objectives. Each exercise is analyzed on an exercise-by-exercise basis to provide a description of students' performance on specific tasks. The purpose of this chapter is to discuss results selected from the second NAEP mathematics assessment that characterize students' mastery of basic skills in mathematics and to identify the implications of these results for instruction in mathematics. For other reports of results of the assessment, see Carpenter et al. (1980a, b, and in press) as well as the National Assessment of Educational Progress (1979a, b, c, d).

DEFINITION OF BASIC SKILLS

In order to describe students' knowledge of basic skills, it is necessary to identify what constitute basic skills. At a superficial level, basic skills are often equated with computation. We have chosen to take a broader view. A number of broadly based definitions of basic skills have been proposed (Commission on Post-War Plans 1947; Committee on Basic Mathematical Competencies and Skills 1972), but none of the proposed definitions has received universal acceptance. One list of basic skills that has received more widespread acceptance than others, however, is that developed by the National Council of Supervisors of Mathematics (NCSM). Although other characteristics might be chosen, the NCSM outline presented in Table 13-1 provides a reasonably comprehensive framework to serve as a basis for our analysis. Because several of the basic skill areas are very closely related, they have been clustered together for purposes of discussion.

The basic skills cited in Table 13-1 are limited to cognitive objectives. An important goal of every mathematics program, however, is to develop not only competence but also a positive attitude toward mathematics. The second mathematics assessment collected extensive data on the affective domain. These data not only represent a measure of an important goal of the mathematics curriculum, they also provide a context in which to interpret the results of the cognitive exercises. The attitudinal data are summarized in the concluding section of this analysis of basic skills in mathematics.

Table 13-1

Highlights of ten basic skill areas

Problem solving: Students should be able to solve problems in situations that are new to them.
Applying mathematics to everyday situations: Students should be able to use mathematics to deal with situations they face daily in an ever-changing world.
Alertness to the reasonableness of results: Students should learn to check to see that their answers to problems are "in the ball park."
Estimation and approximation: Students should learn to estimate quantity, length, distance, weight, and so on.
Appropriate computational skills: Students should be able to use the four basic operations with whole numbers and decimals, and they should be able to do computations with simple fractions and percents.
Geometry: Students should know basic properties of simple geometric figures.
Measurement: Students should be able to measure in both the metric and customary systems.
Tables, charts, and graphs: Students should be able to read, make, and interpret simple tables, charts, and graphs.
Using mathematics to predict: Students should know how mathematics is used to find the likelihood of future events.
Computer literacy: Students should know about the many uses of computers in society, and they should be aware of what computers can do and what they cannot do.

Source: National Council of Supervisors of Mathematics

ASSESSMENT SAMPLE AND PROCEDURES

In order to interpret the results of the National Assessment, it is important to take into account the nature of the sample and certain aspects of the assessment procedure. For a more complete description of procedures, see National Assessment of Educational Progress (1978). Since respondents are selected by age, students from several different grades were included at each age level. Approximately 25 percent of the 9-year-olds were in grade three, and most of the remaining 75 percent were in grade four. Approximately 27 percent of the 13-year-olds were in grade seven, and 69 percent were in grade eight. The 17-year-old sample consisted only of students who were still in school. About 72 percent of these students were in grade eleven, 14 percent in grade ten, and 14 percent in grade twelve.

There was also a great deal of variability within the 17-year-old sample in terms of course background. The percentages of 17-year-olds who had completed at least half a year of selected mathematics

Table 13-2

Mathematics courses taken by 17-year-olds

Course	Percent having completed at least one-half year
General or business mathematics	46
Prealgebra	46
Algebra I	72
Geometry	51
Algebra II	37
Trigonometry	13
Precalculus/calculus	4
Computer programming	5

courses are shown in Table 13-2. Since nearly seven-eighths of the 17-year-olds were in the tenth or eleventh grade, many of these students may take another mathematics course before completing high school. Nevertheless, these data indicate that there is substantial attrition in upper-level mathematics courses.

The 9-year-olds were assessed during January and February 1978, the 13-year-olds during October and November 1977, and the 17-year-olds during March and April 1978. Altogether, approximately 230 exercises were administered to 9-year-olds, 350 to 13-year-olds, and 430 to 17-year-olds. Since testing time was limited to forty-five minutes for each participant, an item-sampling procedure was used in which each exercise was administered to approximately 2,400 respondents at each age level. The item-sampling procedure makes a "total score" inappropriate for an individual; rather, results are reported on an exercise-by-exercise basis in terms of the percentage of respondents at each age level who gave particular responses. Certain exercises are not released so that they can be used to measure change in performance in future assessments. When the unreleased exercises are needed to round out the interpretation of the results reported in this chapter, they have been described in general terms or similar exercises have been constructed to illustrate the nature of the problem.

Both multiple choice and open-ended exercises were included. Scoring guides were developed for the open-ended exercises so that the percent of respondents making specific errors could be identified.

All exercises were administered by specially trained exercise administrators to groups of fewer than twenty-five students. To standardize procedures and minimize reading difficulty, all exercises were presented on a paced audiotape as well as in exercise booklets.

RESULTS

Problem Solving and Application

Problem Solving. Learning to solve problems is the principal reason for studying mathematics. Problem solving is the process of applying previously acquired knowledge to new and unfamiliar situations. Solving word problems in texts is one form of problem solving, but students also should be faced with nontextbook problems. Problem-solving strategies involve proving questions, analyzing situations, translating results, drawing diagrams, and using trial and error. In solving problems, students need to be able to apply the rules of logic necessary to arrive at valid conclusions. They must be able to determine which facts are relevant. They should be unfearful of arriving at tentative conclusions, and they must be willing to subject these conclusions to scrutiny. (NCSM 1978)

Applying Mathematics To Everyday Situations. The use of mathematics is interrelated with all computation activities. Students should be encouraged to take everyday situations, translate them into mathematical expressions, solve the mathematics, and interpret the results in light of the initial situation. (NCSM 1978)

Generally, performance on simple one-step word problems was good. In fact, the difference between performance on a verbal problem and performance on a corresponding whole number computation exercise was almost always less than 10 percent at the upper age levels. For example, 38 percent of the 9-year-olds and 82 percent of the 13-year-olds correctly solved a problem similar to the following:

George had 342 mathematics problems to do for homework. If he did 278 of them in school, how many did he have left to do at home?

For the corresponding computation problem, 342 − 278, the correct answer was calculated by 50 percent of the 9-year-olds and 85 percent of the 13-year-olds.

Although students could successfully identify which operation should be used to solve simple one-step problems, they had great difficulty analyzing nonroutine or multistep problems. In fact, given a problem that required several steps or contained extraneous information, students frequently attempted to apply a single operation to

Table 13-3

Nonroutine word problems

Exercise	Percent Correct	
	Age 13	Age 17
A recipe for punch calls for 3 $\frac{3}{4}$ cups of pineapple juice for ten people. How much pineapple juice should be used to make the same punch to serve five people?		29**
Lemonade costs 95¢ for one 56-ounce bottle. At the school fair, Bob sold cups holding 8 ounces for 20¢ each. How much money did the school make on each bottle?*	11	29

*The actual exercise is unreleased but is similar in content to the problem given here.
**This percent includes those that gave the correct numerical value with or without the correct label.

the numbers given in the problem. Students' difficulty with nonroutine and multistep problems is illustrated by the results summarized in Table 13-3. It is particularly enlightening that in the second exercise almost 40 percent of the 13-year-olds and 25 percent of the 17-year-olds took two of the numbers and performed a single operation to arrive at a solution.

Even when students could identify the appropriate operation, they frequently had difficulty relating the result of their calculation to the given problem in nonroutine situations. For example, 13-year-olds were told that a certain number of items could be packed in a box and were asked how many items would be left over after a given number of items had filled as many boxes as possible. Of the 13-year-olds, 29 percent recognized that the remainder to the division calculation was the correct response, but 22 percent gave the quotient as their answer. This error resulted from the fact that the problem required the students to do more than routinely identify an appropriate operation and perform the calculation. The results indicate that for too many students, the process of problem solving involves little more than that type of routine.

An important aspect of problem solving identified in the NCSM recommendations is identifying which facts are relevant to a given problem. The exercise in Table 13-4 illustrates the difficulty students have dealing with extraneous data. Almost a fourth of the 13-year-olds

Table 13-4

A problem with extraneous data

One rabbit eats 2 pounds of food each week. There are 52 weeks in a year. How much will 5 rabbits eat in one week?

	Percent Responding	
	Age 9	Age 13
0 2 pounds	2	2
● 10 pounds	47	56
0 52 pounds	16	5
0 104 pounds	16	11
0 520 pounds	12	23
0 I don't know	6	3

tried to incorporate all the numbers given in the problem into their calculation. This is another example of the routine application of some operation to all the numbers given in a problem rather than analyzing the problem to determine which data are relevant and planning how to solve the problem.

One of the difficulties students have in solving problems is that they have not developed good problem-solving strategies. A basic strategy that helps in analyzing certain types of problems is to draw a picture. In the pair of exercises in Table 13-5, the same problem was given in one exercise with an accompanying figure and in the other in a problem situation without a figure. This variation produced a difference of over 30 percentage points in the percent of correct responses for both 9- and 13-year-olds. Since most students could identify a rectangle on a simple recognition task, it appears that many of them did not use this knowledge to draw a figure in order to help them solve the verbal problem.

It is virtually impossible to test students' ability to apply mathematics to everyday situations using a paper-and-pencil test. An exercise on a test simply does not represent a real problem for the students as they would experience it outside of school. Therefore, the results of the exercises presented in Table 13-6 should be interpreted with caution.

The results for the exercises in Table 13-6, as well as a number of other exercises involving applications of mathematics, paralleled the results on problem-solving exercises in general. Students were suc-

Table 13-5

Distance problems with and without a picture

What is the *distance all the way around* this rectangle?

		Percent Responding Age 9	Age 13
0	16 feet	39	12
0	30 feet	4	1
●	32 feet	40	60
0	36 feet	4	4
0	60 feet	4	13

Mr. Jones put a wire fence all the way around his rectangular garden. The garden is 10 feet long and 6 feet wide. How many feet of fencing did he use?

		Percent Responding Age 9	Age 13
0	16 feet	59	38
0	30 feet	6	3
●	32 feet	9	31
0	36 feet	5	5
0	60 feet	15	21

cessful on simple applications of arithmetic, like making change, but had more difficulty with problems in which it was necessary to analyze the problem situation to figure out what needed to be done.

It is possible that students are somewhat more successful in solving problems in real-life situations. Most textbook problems can be solved with a relatively routine application of computational skills. It may be that students have learned that most textbook problems do not require careful analysis, and they may be more motivated to think through a problem in a real-life situation. For example, the results for the problem involving cooking a roast suggest that most Americans

Table 13-6

Applied problems

Problem	Percent Correct	
	Age 13	Age 17
Ralph bought a can of tennis balls for $3.78 and a pair of sweat socks for $2.95. He paid for his purchase with a twenty-dollar bill. How much change should he receive?*	62	77
A roast is to be cooked 20 minutes for each pound. If a roast weighing 11 pounds is to be done at 6:00 p.m., what time should it be put in the oven to cook?		25
A quart of asphalt paint covers 200 square feet of surface. How many quarts of paint are needed to paint a driveway 30 feet wide and 160 feet long?*		26

*The actual exercise is unreleased but is similar in content to the problem given here.

would have difficulty planning dinner. However, most cooks do manage to get meals on the table approximately on time. These results perhaps say a great deal more about students' ability to reconcile their everyday experiences with the symbolic problems they encounter in school than they do about students' ability to deal with real-life problems. Nevertheless, such exercises—artificial though they may be—represent one of the few gauges available to assess the applied problem-solving performance of students that results from their formal schooling. These results leave no doubt that there is much room for improvement.

Estimation and Reasonableness

Estimation-Approximation. Students should be able to carry out rapid approximation calculations by first rounding off numbers. They should acquire some simple techniques for estimating quantity, length, distance, and so on. It is also necessary to decide when a particular result is precise enough for the purpose at hand. (NCSM 1978)

Alertness to the Reasonabless of Results. Because of arithmetic errors or other mistakes, the results of mathematical work are sometimes wrong. Students should learn to inspect all results and to check for reasonableness in terms of the original problem. With the increase in the use of calculating devices in society, this skill is essential. (NCSM 1978)

Table 13-7

Subtraction-estimation

$$\begin{array}{r} 6058 \\ -4875 \\ \hline \end{array}$$

The answer to this subtraction problem is closest to:*

Response		Percent Responding Age 9	Age 13	Age 17
●	1000	17	52	69
0	2000	39	34	25
0	3000	16	6	4
0	9000	19	6	1
0	I don't know	8	2	0

*The actual exercise is unreleased but is similar to the problem given here.

Estimation-approximation and alertness to reasonableness of results are closely related. Alertness to reasonableness often involves comparing a result to an estimate, and estimation is basically a process of figuring out what a reasonable result should be. Because of the difficulty in separating these skills in a meaningful fashion, we will discuss them together.

The very nature of estimation makes a formal assessment difficult. When given a computational estimation problem, it is difficult to know the extent to which estimation is actually used. Unless the time allowed for a problem is carefully controlled, a complete computation may actually be performed rather than an estimation. Consequently, the results for the few estimation exercises administered should be interpreted with caution. The available evidence suggests, however, that at all age levels, students are not proficient at estimation.

The results for a computation estimation exercise are presented in Table 13-7. At each age level, performance is over 15 percentage points lower than on a corresponding exercise in which students were asked to compute the exact difference between two numbers. Students were also more successful in performing calculations with fractions and decimals than in estimating the result of the operations. The estimates to the fraction and decimal exercises also gave evidence of a lack of alertness to the reasonableness of results. For example, when

Table 13-8

Performance on computation estimation exercise

ESTIMATE the answer to $\frac{12}{13} + \frac{7}{8}$. You will not have time to solve the problem using paper and pencil.

		Age 13	Age 17
0	1	7	8
●	2	24	37
0	19	28	21
0	21	27	15
0	I don't know	14	18

asked to estimate the sum of three addends, such as 295.0 + 865.2 + 1.583, more than 70 percent of the 13-year-olds and 50 percent of the 17-year-olds chose either 10,000 or 100,000 as the sum.

It is also difficult to separate estimation and alertness to the reasonableness of a result from other mathematical skills. One reason students have difficulty estimating is that they have not mastered the mathematics involved in the estimate. Many of the errors in the exercise in Table 13-8 were based on fundamental misconceptions about fractions. Many of respondents at each age level simply added the numbers in the numerator or denominator, obtaining a completely unreasonable result if one understands what it means to add two fractions, each with a value of less than 1. However, rather than estimating the sum, many students appear to have attempted to find some way to operate directly on some of the numbers given in the problem with no concern for the reasonableness of their work.

The NCSM recommendations indicate the essential nature of alertness to the reasonableness of results given the widespread use of calculators. A reason for this concern is illustrated by the results of an exercise in which students provided with a calculator were asked to perform a calculation similar to 9 ÷ 17. Of the 9-, 13-, and 17-year-olds, 26, 22, and 20 percent, respectively, ignored the decimal point and reported the eight digits displayed on their calculators.

Alertness to reasonableness of a result not only involves deciding whether an answer to a calculation is of the proper order of magnitude, it also involves determining whether the result fits the context of the problem. Consider the following problem:

A boy had 1,570 cans of pop to pack in boxes that hold 24 cans each. How many cans of pop will be left over after the boy has filled as many boxes as he can?

Only a whole number of pop cans would make an appropriate answer and no more than 23 should be left over. Yet, 22 percent of the 13-year-olds working without a calculator responded 65 5/12 or 65.4167, and 8 percent of them reported 5/12 cans were left over. When a calculator was available, 26 percent of the 13-year-olds and 19 percent of the 17-year-olds responded 65.4167 or 65. The fact that these answers were out of the domain of acceptable responses was not taken into account by many of these students.

Two different levels of skill in estimating measurements were also assessed. One was tested using a multiple-choice format that required respondents to select the measure that was close to the height of a common object. These exercises included choices as different as 1 foot, 6 feet, and 15 feet. Essentially, these exercises tested little more than familiarity with the length of the unit of measure. The second type of exercise asked respondents to estimate the length of a segment to the nearest inch or centimeter, allowing an error of one unit. This is a more difficult problem and requires the application of basic estimation skills like partitioning the segment into smaller segments. Both types of exercises were administered with metric and English units. The results summarized in Table 13-9 indicate that few children at any level have mastered basic estimation skills. Although there was a great

Table 13-9

Length estimation exercises

Problem	Percent Correct		
	Age 9	Age 13	Age 17
Gross estimate of the height of a common object in English units.	53	79	86
Gross estimate of the height of a common object in metric units.	20	37	50
Estimate the length of a segment to the nearest inch.	35		
Estimate the length of a segment to the nearest centimeter.	29	30	22

disparity between the results of the metric and English unit problems requiring only a general familiarity with units, there was relatively little difference between the metric and English unit problems requiring estimation to the nearest unit.

Students appear to be able to make gross approximations of the number of objects in a collection. Nine-year-olds were asked to choose whether there were 20, 200, or 2000 birds in the picture in Figure 13-1. Almost three-fourths chose the correct answer of 200. Most of the rest of the 9-year-olds chose 2000 as their estimate.

Computation

Appropriate Computational Skills. Students should gain facility with addition, subtraction, multiplication, and division of whole numbers. Today it must be recognized that long, complicated computation will usually be done with a calculator. Knowledge of single digit number facts is essential and mental arithmetic is a valuable skill. Moreover, there are everyday situations which demand recognition of, and simple computations with, common fractions.

Because consumers continually deal with many situations that involve percentage, the ability to recognize and use percents should be developed and maintained. (NCSM 1978)

Figure 13-1

Measurement estimation exercise

Table 13-10

Average percents on basic facts

Type of facts	Number of parts	Average Percent Correct Age 9	Age 13	Age 17
Addition	6	89	95	97
Subtraction	6	79	93	95
Multiplication	6	60	93	93
Division	5	*	81	89

*Division facts were not administered to 9-year-olds.

Table 13-11

Addition, subtraction, and multiplication computation

	Percent Correct Age 13	Age 17
Exercise 4285 3273 +5125	85	90
Subtract 237 from 504	73	84
671 x402	66	77

Whole Number Computation. The results given in Table 13-10 indicate that students have been successful in learning basic facts. In spite of the fact that drill on addition and subtraction facts diminishes between the ages of 9 and 13, performance increases. The use of facts in contexts other than drill must help to maintain and even increase performance.

Performance for whole number computation was also generally good. About 75 percent of the 9-year-olds could perform simple addition computations that required regrouping, and about 65 percent could solve simple subtraction exercises involving regrouping. Almost all students could make simple calculations involving addition, subtraction, and multiplication, and most of them were successful

Table 13-12
Fraction addition

Exercise	Percent Correct	
	Age 13	Age 17
$\frac{4}{12} + \frac{3}{12}$	74	90
$2\frac{3}{5} + 4\frac{4}{5}$	63	77
$\frac{1}{2} + \frac{1}{3}$	33	66
$\frac{7}{15} + \frac{4}{9}$	39	54

with more difficult calculations (see Table 13-11). The results summarized in Table 13-11 do indicate, however, that some 13- and 17-year-olds continue to experience difficulty with problems involving zero.

Both 13- and 17-year-olds experienced difficulty with division. Only about half of the students in both age groups made the following calculation correctly:

$$28 \overline{)3052}$$

Even relatively straightforward division exercises without zeros in the quotient were missed by about 30 percent of the 13-year-olds and 15 percent of the 17-year-olds. It is interesting to compare these results with performance on identical exercises for which calculators were available. Although a great deal of time and effort is devoted to learning the division algorithm, only about half of the oldest students could perform the more complex calculations successfully. By contrast, with almost no formal instruction in the use of calculators, over 80 percent of the 13-year-olds and 90 percent of the 17-year-olds could do the same exercises using a calculator. This raises serious questions about the time spent drilling on division skills. It may be more productive to develop division concepts and estimation skills and then let students use a calculator.

Fractions and Decimals. The results in Table 13-12 indicate that most students can add simple fractions with common denominators. However, to add fractions with unlike denominators, many employ superficial manipulations. For example, to add $\frac{1}{2}+\frac{1}{3}$, 30 percent of the 13-year-olds and 15 percent of the 17-year-olds simply added the numerators and denominators to get an answer of $\frac{2}{5}$.

It is interesting to note that the complexity of the denominators had relatively little effect on students' ability to add fractions with unlike denominators. It appears that if students have learned a computational algorithm, they can apply it successfully in most situations. However, if they have not mastered the algorithm, they cannot solve even simple problems that might be solved intuitively or by using simple models of fractions.

Performance on subtraction exercises was at approximately the same level as performance on similar addition exercises. Problems that required regrouping or borrowing were significantly more difficult and were only solved by 18 percent of the 13-year-olds and 37 percent of the 17-year-olds.

Slightly more than half of the 13-year-olds and about two-thirds of the 17-year-olds could make calculations like $\frac{5}{8} \times \frac{3}{7}$. Performance dropped to 30 percent and 40 percent, respectively, when a mixed numeral was one of the factors. Division of fractions was not assessed.

Overall, it appears that roughly two-thirds of the 13-year-olds and about three-fourths of the 17-year-olds have learned most of the very elementary skills involving fractions. However, only about half of this number can integrate these skills to solve some of the more involved calculations with unlike denominators and mixed numerals. In other words, only about 40 percent of the 17-year-olds appear to have mastered basic computation with fractions.

If basic decimal concepts have been mastered, then operations with decimals are essentially reduced to operations with whole numbers. About 50 percent of the 13-year-olds and 70 percent of the 17-year-olds were successful in exercises assessing basic decimal concepts, and about the same percentages of each age group could successfully add, subtract, and multiply decimals. Performance on division exercises with a decimal point in the divisor was at about the 33 percent level for 13-year-olds and 50 percent of 17-year-olds.

Percents. Overall performance on percent exercises was extremely low. About 33 percent of the 13-year-olds and 50 percent of the 17-year-olds could express numbers like $\frac{9}{100}$ or .3 as a percent. About the same percent of students could solve problems involving familiar percents like 25% and 50%. For example, 35 percent of the 13-year-olds and 58 percent of the 17-year-olds could find what percent 30 is of 60. Calculations involving unfamiliar percents were much more difficult and were solved by fewer than 10 percent of the 13-year-olds and 30 percent of the 17-year-olds. For example, only 8 percent of the 13-year-olds and 27 percent of the 17-year-olds could find 4% of 75.

In general, students' success with computation appeared to be closely related to how well they had learned the basic number concepts underlying the computation. Students demonstrated a good understanding of whole number concepts, and computation with whole numbers was generally good. Difficulties with basic concepts of fraction, decimal, and percent were paralleled by errors in computation. Frequently, these errors resulted in completely unreasonable results, which students did not recognize because they did not have sufficient understanding of basic number concepts.

It is also important to note that mastery of many skills is not accomplished at the time of major emphasis in the curriculum. The stabilization of these skills comes after they are used for some time and in many contexts. For example, even though there is very little systematic instruction on whole number computation in high school, whole number computation improves about 10 percentage points from age 13 to age 17. Although some skills will continue to develop through use in other contexts, this is not always the case. The current high school curriculum does not take into account that many basic skills are not well developed by the time students begin instruction in algebra and geometry. For example, very few 13- and 17-year-olds have mastered percent concepts or skills. But outside of general mathematics, there is very little opportunity for high school students to extend or even maintain their knowledge of percent.

Geometry

Geometry. Students should learn the geometric concepts they will need to function effectively in the 3-dimensional world. They should have knowledge of concepts such as point, line, plane, parallel, and perpendicular. They should know basic properties which relate to measurement and problem-solving skills. They also must be able to recognize similarities and differences among objects. (NCSM 1978)

This description of geometry as a basic skill area encompasses three major categories: (a) knowledge of basic geometric concepts, (b) knowledge of basic properties, and (c) knowledge of relationships among geometric objects. These categories reflect topics that are taught at the

Table 13-13

Geometric recognition tasks

Figure	Percent Correct Age 9	Age 13	Age 17
Triangle	88	85	86
Square	93	96	96
Rectangle	84	92	—
Circle	99	—	—
Cylinder	41	78	93
Cube	85	96	99
Sphere	20	65	79
Geometric plane	21	50	73
Parallel lines	57	90	94
Perpendicular lines	15	33	70
Radius of circle	—	57	78
Chord of circle	—	33	55
Tangent of circle	—	35	59

elementary and junior high levels in informal geometry and are not necessarily relegated to courses in formal geometry at the high school level. Most of the geometry exercises administered in this assessment also reflect concepts commonly presented in the context of informal geometry.

Table 13-13 summarizes the results of several exercises that required recognition of simple geometric figures. The table shows that most respondents were successful on most of the exercises. Apparently, school instruction and real-life situations together provide the kinds of experience students need to become familiar with basic geometric concepts. As expected, there was a noticeable difference between recognition and use of technical terms (sphere, perpendicular) and more common everyday terms (cube, parallel).

Levels of performance on exercises that dealt with basic properties of geometric figures were much lower overall than those reported for knowledge of geometric concepts. In fact, with rare exceptions, students were generally unfamiliar with all but the most fundamental properties. For example, four exercises were administered that assessed knowledge of the property that the sum of the measures of the interior angles of a triangle is 180 degrees. Success rates on these exercises for the 17-year-olds ranged from 40 to 60 percent correct. On these and similar exercises, however, the taking of a course in geometry made significant difference in performance. For the exercises described, success rates for those 17-year-olds with a year of geometry ranged from 68 to 86 percent correct; for those 17-year-olds with no geometry course, percents correct were from 9 to 32. Thus, even though the geometric concepts assessed were not necessarily restricted to a formal course in geometry, the experience provided by a formal course apparently made a difference in levels of performance.

Although most of the exercises that dealt with the properties of simple figures were administered to the older respondents, some exercises were given to the 9-year-olds. For example, two-thirds of the 9-year-olds were able to determine the length of a side of a square from a diagram even though that particular side was not labeled. Over half and almost two-thirds of the 9-year-olds and 13-year-olds, respectively, could select a triple of line segments whose union would not form a triangle. The nature of the exercise and the language used to state the question were factors that were seen to influence performance.

Language factors also influenced performance on some exercises that assessed students' knowledge and understanding of relationships among classes of geometric objects. For example, 82 and 94 percent of the 9-year-olds and 13-year-olds, respectively, were able to select a pair of figures that had the "same size and shape," but performance levels dropped on comparable exercises that used the term "congruent." This same tendency was observed on exercises that employed the concept of similar figures. The students were also successful on exercises that required recognition of similar figures from a diagram but were much less successful at dealing with the properties of those figures in problem situations.

An exercise that dealt directly with 17-year-olds' knowledge of relationships among certain types of quadrilaterals is shown in Table 13-14. The results of this exercise suggest that a majority of students know that squares and rectangles are types of parallelograms (statements B and D) and consequently, that it is not true that every parallelogram is a square (statement A). They do not, however, recognize the relationship between squares and rectangles (statement C).

Properties of relationships that would be useful in problem-solving situations, such as the Pythagorean Theorem, were also assessed; how-

Table 13-14

Results on an exercise about relationships among quadrilaterals

Statement	Percentage Responding		
	True	False	I don't know
A. Every parallelogram is a square.	16	73*	11
B. Every square is a parallelogram.	62*	27	11
C. Every square is a rectangle.	32*	65	2
D. Every rectangle is a parallelogram.	64*	25	10

*Indicates correct response

ever, students were relatively unsuccessful in being able to recognize and apply such properties in appropriate situations. Performance levels for the 17-year-olds on four exercises that required application of the Pythagorean Theorem ranged from 8 to 49 percent. Once again, course background made substantial differences.

If the results obtained on the geometry exercises are weighed against the NCSM basic skills in geometry, a mixed picture emerges. Students appear to have knowledge of basic geometric concepts. On the other hand, students are not generally able to deal with the properties of simple geometric figures or to analyze relationships among classes of figures, and they are not able to apply their knowledge in problem situations, a trend that has been observed across all of the skill areas.

Measurement

Measurement. As a minimum skill, students should be able to measure distance, weight, time, capacity, and temperature. Measurement of angles and calculations of simple areas and volumes are also essential. Students should be able to perform measurement in both metric and customary systems using the appropriate tools. (NCSM 1978).

Since the assessment relied primarily on a paper-and-pencil format, there was little opportunity to assess directly students' ability to measure weight or capacity. But on certain exercises, they were provided rulers, and their ability to read thermometers, scales, and clocks was assessed using pictures of the different instruments.

Measuring Length. At all three ages, most students could make simple measurements of length. Eighty-one percent of the 9-year-olds and

How long is this line segment?

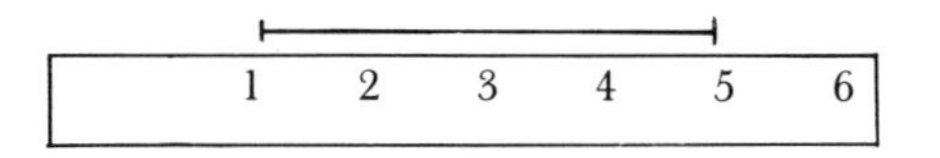

Figure 13-2

Illustration of a linear measurement exercise

over 90 percent of the 13- and 17-year-olds could measure the length of a segment that measured a whole number of units. Measurements that involved fractions of inches or required the addition of several linear measures were significantly more difficult. Fewer than two-thirds of the 13-year-olds and about three-fourths of the 17-year-olds could make these more complex measures.

Although most students were able to use a rule to make a simple linear measurement, an exercise similar to the one in Figure 13-2 is some indication that their understanding of this skill may be superficial. Only 19 percent of the 9-year-olds and 59 percent of the 13-year-olds answered this question correctly. Seventy-seven percent and 40 percent, respectively, gave an answer of 5, completely ignoring the fact that the end points were not aligned. This is an example in which a slight change in context exposed a superficial understanding on the part of many students.

Weight, Temperature, and Time. Most 9-year-olds could read a bathroom scale that was marked at each pound but only labeled every 5 pounds even though the indicated weight was not a multiple of five. They had more difficulty reading temperatures when only even degrees were marked on the thermometer. Three-fourths of the 9-year-olds, over half of the 13-year-olds, and over a third of the 17-year-olds incorrectly assumed that each mark represented an interval of one degree.

Students' difficulty in reading thermometers appears to result primarily from characteristics of the scale on the thermometer rather than from the fact that it is inherently more difficult to measure temperature than weight. In fact, one of the unique characteristics of measuring temperature, the fact that there are below-zero temperatures, caused relatively little difficulty. About 80 percent of 9-year-olds and 90 percent of the 13- and 17-year-olds could read a temperature of –10°.

Most 9-year-olds could read a clock, although performance declined with finer gradations of time. Over 85 percent of the 9-year-olds could tell time at fifteen-minute intervals (8:15, 6:45), and almost 60 percent could tell time at one-minute intervals (2:53). By age 13, 85 percent could tell time at one-minute intervals.

Area and Volume. At all ages performance on area and volume skills has not reached desired levels. A basic concept is that area is the number of units, usually square units, required to cover a given region. Calculations of areas by multiplying various linear dimensions of figures should be based on understanding that such operations are a shortcut for finding the number of units in a unit covering. Few 9-year-olds have any knowledge of the basic area concept as illustrated by only 28 percent being able to find the area of a rectangle that was divided into square units.

By age 13, most students can count the number of units in a unit covering, but there is still a substantial number who have not mastered even this basic concept. Although 71 percent of the 13-year-olds could find the area of a rectangle that was divided into square units, only 51 percent could calculate the area of that rectangle from the dimensions of the sides. Few 13-year-olds could find the area of more complex figures. Only 4 percent could find the area of a right triangle, and only 12 percent could find the area of a square given one of its sides.

About three-fourths of the 17-year-olds could calculate the area of a simple rectangle, but only about 40 percent could find the area of a square, and fewer than 20 percent could find the area of a right triangle or parallelogram. In fact, only 34 percent of the 17-year-olds who had completed a full year of geometry could find the area of a right triangle, and only 68 percent could find the area of a square.

As might be expected, students' knowledge of volume concepts was even poorer than their knowledge of area concepts. For example, only 17 percent of the 13-year-olds and 39 percent of the 17-year-olds could find the volume of a rectangular solid; 42 and 18 percent, respectively, added the three dimensions given for the solid. Only 57 percent of the 17-year-olds who had studied a full year of geometry correctly calculated the volume of the solid.

Even the students who had learned basic area and volume formulas had difficulty applying them to problems that required anything more than substituting numbers into a formula. For example, only 16 percent of the 17-year-olds could find the area of a region made up of two rectangles, and only 9 percent could solve the following exercise:

> *How many cubic feet of concrete would be needed to pave an area 30 feet long and 20 feet wide with a layer four inches thick?*

Standard Units of Measure. Students are still more familiar with common English units than they are with metric units, but the gap seems to be narrowing. The greatest improvement over the first assessment on any exercises was on exercises assessing students' familiarity with common metric units. On one exercise given to 13-year-olds, there was a gain of 26 percentage points; and on two exercises given to 17-year-olds, there were gains of 12 and 14 points. By the age of 13, most students have some knowledge of common metric units. About half the 9-year-olds and between 80 and 90 percent of the 13- and 17-year-olds could identify the centimeter as the appropriate unit to measure their thumb. However, although 72 percent of the 17-year-olds could identify which English unit was closest in length to a common metric unit, most students had difficulty making estimates in metric units. Only 20 percent of the 9-year-olds, 37 percent of the 13-year-olds, and 50 percent of the 17-year-olds were able to make even reasonable estimates of weight and length in metric units. At each age, these percentages are about 35 points below the performance level for corresponding problems involving English units. Thus, although it appears that most students have had some exposure to metric units, most of them have not learned to think in those terms.

Organizing and Representing Data

Tables, Charts, and Graphs. Students should know how to read and draw conclusions from simple tables, maps, charts, and graphs. They should be able to condense numerical information into more manageable or meaningful terms by setting up simple tables, charts, and graphs.

For all age groups, performance on graph or table exercises requiring direct reading of graphs or tables or comparisons from graphs or tables was consistently higher than on those requiring interpolation, extrapolation, or problem solving. This is not surprising, since these

Table 13-15

Results on a graphing exercise

The Steelers football team voted for the most valuable player in the play-offs. Here is how the vote came out.*

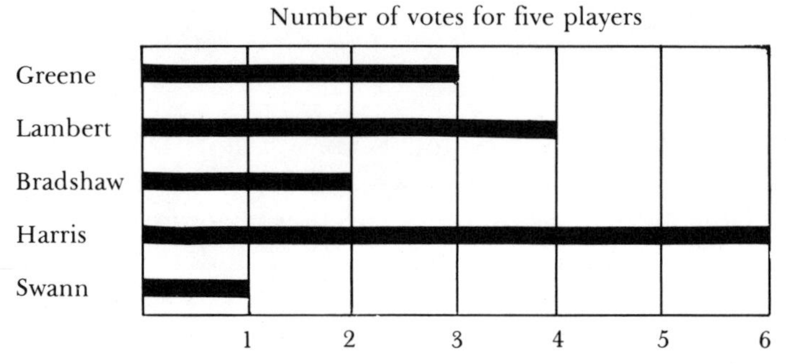

Question	Percent Correct Age 9	Percent Correct Age 13
Who had the most votes?	94	99
Who had the least votes?	90	98
How many votes did Lambert get?	91	98
How many members of the team voted?	41	68

*The actual exercise is unreleased but is similar to the problem given here.

latter tasks clearly require a higher level of understanding and application. These results imply that students need experience not only in building and reading graphs, but also in using them to make comparisons and predictions, and to search for trends and patterns within data.

Students' success with simple graph-reading tasks and their difficulty in answering questions that cannot be read directly from the graph are illustrated by the results in Table 13-15. Forty-two percent of the 9-year-olds and 24 percent of the 13-year-olds responded that a total of exactly six players had voted. It appears that these students were attempting to answer all questions by directly reading the graph and did not realize that some additional calculation was required.

Students often made superficial interpretations of numbers presented in a table or graph, causing them particular difficulty when symbols represented more than a single unit. For example, Table 13-16 was presented to 13- and 17-year-olds. Students were asked how

Table 13-16

Airline passengers for first six months of the year

Airports	Hundreds of Passengers per Month						Total
	Jan	Feb	Mar	Apr	May	June	
Bay City	9	3	5	7	2	4	30
Camden	6	8	1	5	8	2	30
Dover	8	5	9	6	6	3	37
Fiske	5	6	6	1	3	7	28
Grange	1	2	3	6	7	10	29
Total	29	24	24	25	26	26	154

many passengers used the Fiske Airport in June. Only 18 and 32 percent of the 13- and 17-year-olds, respectively, answered this multiple-choice question correctly; 60 percent of the 13-year-olds and 50 percent of the 17-year-olds responded with the answer "7." Although they were correctly reading the numerical value reported, other important information reported in the table was neglected by the students.

Several exercises presented table-reading problems similar to those students might encounter outside of school. One was an exercise that asked 13- and 17-year-olds to determine the amount of sales tax for specified amounts using a tax chart similar to that in Figure 13-3.

Sale	Tax
$ .00– .12	$ .00
.13– .36	.01
.37– .60	.02
.61– .85	.03
.86–1.09	.04
1.10–1.33	.05
1.34–1.57	.06
1.58–1.82	.07
1.83–2.06	.08
2.07–2.30	.09
2.31–2.54	.10
2.55–2.79	.11
2.80–3.03	.12

Figure 13-3

Tax collection schedule

Slightly less than 50 percent of the 13-year-olds and about 75 percent of the 17-year-olds could complete this task successfully. One portion of this exercise asked students to determine the tax on an amount that exceeded the largest value in the table. As might be expected, extremely low performance resulted, with only 5 percent of 13-year-olds and 27 percent of 17-year-olds answering this open-ended exercise corrrectly.

Another exercise asked 13- and 17-year-olds to use a typical mileage chart to determine the distance between two cities. Fifty-eight percent of the 13-year-olds and 65 percent of the 17-year-olds correctly identified the mileage required.

In general, these results imply that graph-reading skills should not be taken for granted. Although students can read simple graphs and tables, they may ignore important information in the graph or table and be unaware of the importance of the title of the graph, what each axis represents, the key for interpreting symbols, and so forth. They may also encounter difficulty with more complex tables. For example, the mileage chart on which information is organized along several dimensions appeared to cause difficulty.

Probability

Using Mathematics to Predict. Students should learn how elementary notions of probability are used to determine the likelihood of future events. They should learn to identify situations when immediate past experience does not affect the likelihood of future events. They should become familiar with how mathematics is used to help make predictions such as election forecasts. (NCSM 1978)

There are many different ways of using mathematics to predict, and the assessment provided limited evidence of performance on several of them. Several specific exercises that address some important aspects of probability are examined with the hope that these exercises demonstrate the extent to which this basic skill area is being mastered.

Computing the probability of a simple event is the basis for understanding how to determine the likelihood of a future event. Here is an open-ended exercise designed to assess the development of this notion:

2, 3, 4, 4, 5, 6, 8, 8, 9, 10

For a party game, each number shown above was painted on a different Ping-Pong ball, and the balls were thoroughly mixed up in a bowl. If a ball is picked from the bowl by a blindfolded person, what is the probability that the ball will have a 4 on it?

This is a classic probability question with the sample space (ten Ping-Pong balls) and the number of successful outcomes (two balls with a 4) clearly identified. Only about one-third of either 13- or 17-year-olds responded correctly. Performance was even poorer on exercises in which respondents had to construct the sample space themselves. For example, only 5 percent of the 17-year-olds could determine the probability of three coins all coming up heads on a single simultaneous toss. A majority of both 13- and 17-year-olds felt that immediate past experience would affect the likelihood of future events. For example, suppose a fair coin is tossed and four consecutive heads have occurred. What is the probability of getting a head on the next toss? Only about 20 percent of the 13-year-olds and 40 percent of the 17-year-olds answered the question correctly.

Tossing coins, rolling dice, and drawing cards from a deck provide classical cases of independent events. However, it is interesting to examine a less typical probability situation involving independent events. For example, Table 13-17 reports results on an exercise given to 17-year-olds. It appears that most respondents simply ignored the data given in this exercise and appealed to their common sense to conclude that the probability of a boy or girl is equal.

Table 13-17

Performance on a prediction exercise

In the United States, of every 1,000 babies born, 515 are boys. In a certain U.S. hospital, the last 27 babies born have been girls.

The next baby in the hospital will

		Age 17
0	almost certainly be a girl (over 80% chance).	5
0	almost certainly be a boy (over 80% chance).	14
●	have a slightly greater chance of being a boy than a girl.	30
0	have a slightly greater chance of being a girl than a boy.	8
0	have an equal chance of being a girl or boy.	38
0	I don't know.	5

Figure 13-4 presents some data based on 100 samples (that is, spins). Students are asked to predict which spinner was involved. After analyzing the information, students should realize that such a distribution probably resulted from a spinner that is most likely to land in

Kim spun a spinner 100 times and made a record of her results.

Outcome	A	B	C
Number of times	55	30	15

Which spinner is most likely the one that Kim used? Fill in the oval beside the one you choose.

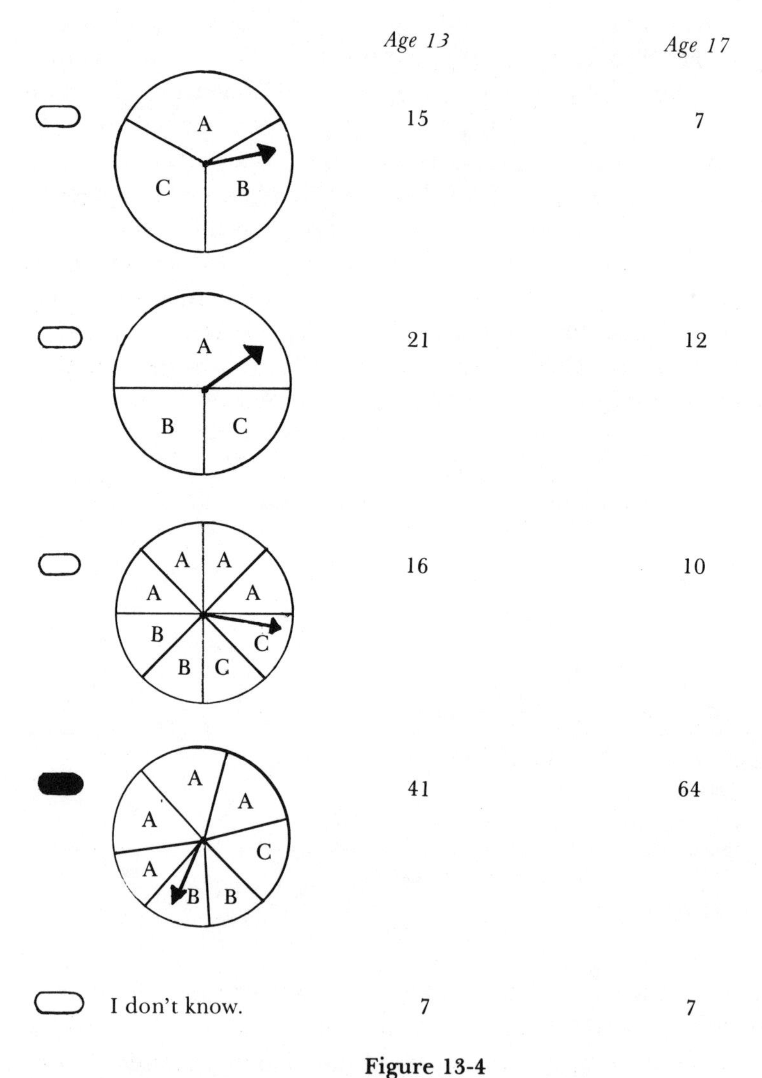

	Age 13	*Age 17*
Spinner 1	15	7
Spinner 2	21	12
Spinner 3	16	10
Spinner 4 (filled)	41	64
I don't know.	7	7

Figure 13-4

Using probability to make decisions

A and least likely to land in C. About two-fifths and two-thirds of the 13- and 17-year-olds, respectively, correctly identified the spinner most likely used. These results suggest that associating a probability model with specific data is a skill in need of further development. In fact, these results are generally representative of the performance level both age groups reached on exercises when data were provided and some decisions requiring predictions were necessary.

Other facets of probability are important, and some of these—such as expected values—were assessed. There was, however, no evidence in the results of the probability exercises to suggest that performance levels of either 13- or 17-year-olds were satisfactory. If using mathematics to predict is a basic skill for schools to promote, and if the current performance is used as a gauge of the progress toward this goal, then much work remains to be done.

Computer Literacy

Computer Literacy. It is important for all citizens to understand what computers can and cannot do. Students should be aware of the many uses of computers in society, such as their use in teaching/learning, financial transactions, and information storage and retrieval. The "mystique" surrounding computers is disturbing and can put persons with no understanding of computers at a disadvantage. The increasing use of computers by government, industry, and business demands an awareness of computer uses and limitations. (NCSM 1978)

Several exercises were included in the mathematics assessment that ascertain 13- and 17-year-old students' awareness of the potential uses of computers, their beliefs about the impact of computers upon society, and other commonly held beliefs about computers.

Students held fairly definite opinions about the potential usefulness of computers in various situations as shown in Table 13-18. More 17-year-olds than 13-year-olds thought a computer would help "a lot" on performing tasks (statements A, B, and E) that are representative of activities frequently accomplished by computers. Both age groups were in agreement that computers would not be very useful in deciding which food tastes best or in writing a novel. Opinion was divided with respect to the perceived usefulness of computers in the prediction of election results or the diagnosis of diseases.

A fourteen-part exercise asked students to respond along a five-point scale from "strongly disagree" to "strongly agree" on statements that reflected specific beliefs and feelings about computers. (For ease of discussion, the two extreme responses will be reported with the

Table 13-18

Results on a computer-awareness exercise

How much would a computer help in:	Age	Percent Responding		
		Computer helps a lot	Helps a little	Does not help
A. Calculating and printing bills for a credit card company	13 17	72 84	24 14	3 1
B. Keeping track of a carmaker's inventory	13 17	48 60	42 35	10 5
C. Telling which food tastes best	13 17	3 1	6 6	90 93
D. Predicting election results	13 17	41 43	35 42	24 16
E. Alphabetizing lists of names	13 17	66 79	27 16	7 5
F. Writing a novel	13 17	7 2	18 11	75 87
G. Diagnosing diseases	13 17	25 24	33 43	42 33

disagree or agree categories). Students were knowledgeable about the simple mechanics of computer operation; over 85 percent of both age groups knew that computers had to be programmed to follow precise instructions and to store information. Around two-thirds of both groups recognized that a computer operator was not necessarily a mathematician.

What kinds of tasks are computers best suited to perform? Fourteen and 11 percent of the 13- and 17-year-olds, respectively, disagreed with the statement "Computers are suited for doing repetitive, monotonous tasks" whereas 39 percent of the 13-year-olds and 63 percent of the 17-year-olds agreed. The 17-year-olds were consistent with their response to a similar statement: "Computers have a mind of their own." On this statement, 67 percent of the 17-year-olds disagreed,

Table 13-19

Responses to statements of feelings about computers

Statement	Age	Percent Responding		
		Disagree	Undecided	Agree
A. Computers dehumanize society by treating everyone as a number	13 17	31 31	37 28	32 41
B. The more computers are used, the less privacy a person will have	13 17	42 40	22 24	35 35
C. Someday most things will be run by computers	13 17	10 9	15 12	76 79

and 18 percent agreed. This consistency was not as strong with 13-year-olds.

The statements presented in Table 13-19 represent students' value judgments about the possible impact of computers upon society. Results on statements A and B of the exercise showed agreement between the two age groups in their responses and a great deal of divided opinion within the groups. Perhaps the most surprising results were obtained on statement C. Over three-fourths of both age groups agreed that computers were ultimately going to be responsible for the operation of almost everything. Although most students in both age groups felt it inevitable that "someday most things will be run by computers," their responses to other statements indicated that they did not feel particularly threatened by that possibility.

The influence of computers upon the job market was reflected in other statements. Around half of both age groups felt that "computers will probably create as many jobs as they eliminate," and slightly more than half of both groups agreed that having a knowledge of computers would help an individual get a better job. A majority of respondents also believed that computers could help make mathematics more interesting. About one-fourth of both age groups were undecided on that issue, which is not at all surprising considering only 14 and 12 percent of the 13- and 17-year-olds, respectively, indicated they had studied mathematics through computer instruction.

Seventy-one and 62 percent of the 13- and 17-year-olds thought that computers would be useful in teaching mathematics, although only 12 and 25 percent had access to computer terminals for learning mathematics in their schools. Not only do the results indicate that most 13- and 17-year-old students have not used computers in their mathematics classes; most have had little actual experience with computers. Only 8 and 13 percent of the 13-year-olds and 17-year-olds said they knew how to program a computer, but it is noteworthy that around two-thirds of both groups thought computer programming would be a useful topic to study in mathematics classes.

Overall, the results indicate that most 13- and 17-year-old students have had little firsthand experience with using or programming computers either inside or outside the mathematics classroom. Despite their lack of experience, most students had some knowledge about the capabilities and limitations of computers. Whether or not the patterns of responses are indicative of acceptable levels of computer literacy is difficult to ascertain. Since so few students were actually knowledgeable about computer operation, the students may have been responding on the basis of their own perceptions of computer mystique rather than from any carefully reasoned point of view. Since this may have been the case, any conclusions about levels of computer literacy among students would at best be tenuous. Nevertheless it is significant that a majority of both age groups not only recognized the importance of computers in their lives, but also indicated that topics related to computers should be integrated into current mathematics classes.

Attitudes

The attitude exercises administered by the National Assessment were grouped into four categories that reflected students' experiences with mathematics in several settings: (a) mathematics in school; (b) mathematics and oneself; (c) mathematics and society; and (d) mathematics as a discipline.

Mathematics in School. This category contained items that ascertained students' feelings about mathematics as a school subject in comparison to other school subjects, particular items of mathematics content (for example, working with fractions), and specific classroom activities (for example, using a mathematics textbook).

On the school-subject comparison, students were asked to indicate how much they liked or disliked each subject and how easy or hard the subject was for them. The older respondents were also asked to rate the subjects according to their perceived importance. Among the subjects, mathematics was rated important or very important by around 90 percent of the 13- and 17-year-olds. None of the other subjects (science, social studies, English, physical education) rated as high, although English was a close second.

Physical education was the best-liked subject in all age groups (liked by around 80 percent of each group). For the 9-year-olds, mathematics was the second favorite subject (liked by 65 percent), but mathematics came in third for the 13-year-olds (liked by 69 percent, which was fewer than the number who liked science), and last for the 17-year-olds (liked by 59 percent and disliked by 32 percent).

On the easy-hard dimension, 42 percent of the 9-year-olds thought mathematics was easy, 42 percent rated it "in between," and 13 percent said it was hard. An interesting phenomenon occurred in the ratings of mathematics made by the older respondents. Mathematics was rated easy by more of the 13-year-olds (56 percent) and hard by more of the 17-year-olds (48 percent) than any other subject. In fact, mathematics was the only subject for which the pattern of responses from age 13 to age 17 along this dimension was different.

The respondents were also presented with a list of classroom activities and asked to rate them according to how often they had done the activity in their mathematics classes, how much they liked the activity, and how useful they found the activity. The activities may be classified as student-centered, classmate-centered, or teacher-centered; a few of the activities do not fit neatly into this scheme and are described simply as "other classroom activities."

Among the student-centered activities, which included taking mathematics tests, doing mathematics homework, working mathematics problems at the board and alone, and using a mathematics textbook, the activity rated as occurring most often was that of using a mathematics textbook (rated as occurring often by 75 percent of the 9-year-olds, 81 percent of the 13-year-olds, and 87 percent of the 17-year-olds). Working mathematics problems alone was cited as occurring often by around three-fourths of all age groups, with the other activities rated as occurring often or sometimes by a majority of the respondents.

A majority of all age groups indicated that they liked working mathematics problems at the board, working problems alone, and using a textbook; doing mathematics homework was the activity most disliked (by 20, 38, and 49 percent of the 9-, 13-, and 17-year-olds, re-

Table 13-20

Students' indicated frequency of selected classroom activities

	Activity	Age	Percent Responding: Often	Sometimes	Never
Classmate-centered	Helping a classmate do mathematics	9	14	56	29
		13	11	71	17
		17	15	72	13
	Getting help from a classmate on mathematics	9	9	61	30
		13	8	74	17
		17	17	74	8
	Discussing mathematics in class	9	49	38	12
		13	58	38	4
		17	50	43	7
	Working mathematics problems in small groups	9	16	49	35
		13	9	47	44
		17	12	59	28
Teacher-centered	Listening to the teacher explain a mathematics lesson	9	85	11	3
		13	81	16	2
		17	78	19	2
	Watching the teacher work mathematics problems on the board	9	78	18	4
		13	76	21	2
		17	79	18	3
	Getting individual help from the teacher on your mathematics	9	21	67	11
		13	17	71	10
		17	18	70	11

spectively). It may be that the students, particularly the younger ones, were somewhat hesitant to express dislike for any of the activities. For example, 46, 49, and 35 percent of the 9-, 13-, and 17-year-olds, respectively, said they liked taking mathematics tests, with 8, 21, and 30 percent, respectively, expressing dislike.

Table 13-20 summarizes the results for the classmate- and teacher-centered activities in terms of their frequencies of occurrence. Indicated frequencies appear to be fairly consistent across age groups, except for the activity of "getting help from a classmate in mathematics," which apparently occurs more often in mathematics classes of older students. The results obtained on this portion of the exercises suggest that students perceive their role in the mathematics classroom to be primarily passive: they feel that they spend a lot of time listening to the teacher explain mathematics, a lot of time watching the teacher work problems, and a lot of time working problems from the textbook on an individual basis. Most students indicated that they had never made reports or done projects in mathematics, nor had they participated in laboratory-type activities in mathematics classes. About 20 percent of the older students indicated, however, that they had been given some choice about what mathematics to study as compared to 3 percent of 13-year-olds who said so. The figures probably reflect that 17-year-olds choose whether or not they will even take mathematics as a school subject, while 13-year-olds are not given that choice.

The 9-year-olds were presented with six computation activities and four activities that dealt with geometry and measurement topics and were asked to rate them along three-point dimensions of easy–hard, like–do not like, and important–not important. Nine of the ten topics were rated important by at least 70 percent of the respondents; learning about circles, triangles, and other shapes was seen as important by only 52 percent. Among the computation activities (doing addition problems, doing subtraction problems with borrowing, checking a subtraction problem by adding, learning multiplication or times tables, dividing one number by another, and solving mathematics word problems), adding and checking subtraction were rated easy by around 66 percent of the respondents, with easy ratings for other topics ranging down to 31 percent for solving word problems. Learning multiplication tables was the topic selected most often as hard (23 percent). Of the geometry and measurement activities, weighing objects with a scale received the highest percentage of hard ratings (11 percent); learning about circles, triangles, and other shapes was seen as the easiest topic (rated easy by 82 percent). The liking ratings showed surprisingly little variation across the ten topics with ratings ranging from a high of 68 percent for learning about money to a low of 45 percent for solving word problems. Subtraction with borrowing, dividing, and word problems was disliked by around one-fifth of the respondents.

With the 13- and 17-year-olds, the seventeen content topics may be divided into categories of computation, geometry and measurement, estimation, and "others." The computation topics included working with whole numbers, fractions, decimals, and percents and a separate item of doing long division. Long division was rated important by around 70 percent of both age groups. Working with whole numbers was rated easy by the largest percentage of both groups (75 and 81 percent of 13-year-olds and 17-year-olds, respectively) and was the best liked of the computation topics (liked by 64 and 66 percent respectively). Across the other computation topics, easy ratings averaged about 50 percent, with liking ratings slightly lower, although relative rankings on ease and liking were consistent in both age groups.

Table 13-21 summarizes the percentages of important, easy, and hard ratings on the remaining content topics. As Table 13-21 shows, ratings on the importance dimension were generally high for all topics except for the topic of doing proofs. Checking answers to problems, estimating measurement, and solving equations were the topics rated highest in importance.

On the easy dimension, about 25 percent of both age groups were undecided as to whether the geometry/measurement topics and the estimation topics were easy or hard. Doing proofs and programming computers received the highest percentages of undecided votes of all topics along this dimension (an average of over 33 percent of each age group). Working with metric measures received the highest percentages of undecided votes of all topics along this dimension (an average of over 33 percent of each age group). Working with metric measures received the highest percentages of hard ratings in both age groups (34 and 43 percent of the 13- and 17-year-olds, respectively); checking answers received the highest percentages of easy ratings in both groups. The largest differences in ratings between the 13- and 17-year-olds occurred on the topics of solving word problems, with over twice as many of the older respondents rating the topic hard (31 percent compared to 14 percent).

The topic of solving word problems created the largest differences between age groups in ratings along the liking dimension as well. Although around one-fourth of each group said they were undecided as to whether they liked or disliked solving word problems, about twice as many 13-year-olds as 17-year-olds liked doing so, with the proportion reversed on the dislike ratings (23 percent compared to 40 percent, respectively). It may be that part of this increase in dislike for word problems from age 13 to age 17 is due to experiences in solving word problems in algebra, an area that is notorious among students

Table 13-21

Ratings on selected content topics

	Topic	Age	Percent Responding: Important	Easy	Like
Geometry/Measurement	Learning about geometric figures	13	63	32	33
		17	56	35	29
	Measuring lengths, weights, or volumes	13	75	45	40
		17	83	50	40
	Working with metric measures	13	73	34	36
		17	73	31	28
Estimation	Estimating answers to problems	13	63	50	44
		17	66	48	37
	Estimating measurements (lengths, weights, areas, and so on)	13	80	50	43
		17	85	49	38
Other	Solving word problems	13	74	57	59
		17	69	37	32
	Doing proofs	13	48	29	26
		17	43	26	19
	Solving equations	13	82	59	56
		17	80	54	46
	Memorizing rules and formulas	13	69	34	31
		17	57	28	20
	Using charts and graphs	13	74	61	55
		17	71	63	52
	Checking the answer to a problem by going back over it	13	81	68	40
		17	89	76	43
	Programming a computer	13	63	19	30
		17	60	15	25

for being difficult. Evidence from the easy ratings lends some support to this assertion.

If it is assumed that the relatively large percentages of "no response" (averaging around 12 percent on liking and 10 percent on easy) among the 13-year-olds would be evenly distributed across all response options, then the results suggest that across all topics (except checking answers and working with whole numbers, which received equal percentages of dislike ratings from both age groups), more 17-year-olds than 13-year-olds disliked the topics. These results reflect the general decline from age 13 to age 17 in positive feelings toward mathematics reported earlier.

The results suggest some observations. First of all, students tend to view all of the topics they have studied in mathematics as important, and this importance rating is consistent across the two age groups. Second, the younger students tend to view topics as easy or hard regardless of their performance on corresponding cognitive exercises. For example, overall levels of performance on word problems and percentage exercises were fairly low, and yet the respondents apparently did not view these topics as particularly difficult. The older respondents, however, did seem to make more distinction between their actual performance and their feelings about the ease or difficulty of a topic. Third, the perceived ease or difficulty of a topic seemed to correspond fairly consistently with the expressed like or dislike for the topic, although there were some exceptions, most clearly that of checking answers. Perhaps the most striking observation to be drawn from these data is that students are capable of evaluating specific topics within mathematics along several dimensions (for example, importance, ease, liking), and they do in fact make such evaluations. Further, their evaluations may or may not correspond to their teachers' a priori expectations.

Mathematics and Oneself. Likert-type statements were used to measure various affective responses to mathematics such as students' achievement, motivation, and self-concept in mathematics. The achievement motivation statements included the following three items: (1) I really want to do well in mathematics; (2) My parents really want me to do well in mathematics; and (3) I am willing to work hard to do well in mathematics. A large majority of all age groups wished to do well in mathematics, as is evidenced by the 2, 1, and 4 percent of the 9-, 13-, and 17-year-olds who disagreed with the statement, and correspondingly large percentages of the students who indicated their willingness to work hard enough to do well in the subject. Around 90 percent of all age groups perceived that their parents also wanted them to perform well in mathematics.

Two of the self-concept statements showed that a majority of all age groups felt that they were fairly good mathematics students. Fifty-five percent of the 9-year-olds felt that they were good at working with numbers, and an additional 40 percent thought they were sometimes good at the task. Sixty-five and 54 percent of the 13- and 17-year-olds, respectively, felt they were good at mathematics, although 9 and 22 percent felt that they were not. In fact, 69 percent of the 13-year-olds and 53 percent of the 17-year-olds responded that they enjoyed mathematics.

The results obtained from these exercises suggest that across all age groups, a majority of mathematics students perceive themselves as competent, motivated, and enjoying their study of mathematics, although there is a slight decline in overall favorableness toward mathematics from age 13 to age 17.

Mathematics and Society. These exercises dealt with students' perception of the usefulness of mathematics to themselves as individuals and to broader societal concerns. The exercises focused on the general usefulness of mathematics and the job-related importance of mathematics.

Over 75 percent of the 13- and 17-year-olds and 66 percent of the 9-year-olds felt that mathematics was useful in helping solve everyday problems. Further, around 80 percent of the older respondents thought that most mathematics had some practical use. Students were consistent in their belief that mathematics was useful; a large majority of all age groups felt that they could not get along very well in everyday life without using some mathematics.

Most students felt that some knowledge of mathematics was important if a person was to get a good job. Over 80 percent of all three age groups thought a knowledge of arithmetic was important; 72 percent of the 13-year-olds but only 46 percent of the 17-year-olds thought knowledge of algebra and geometry important for the job-seeker. A conclusion that might follow is that experience with algebra and geometry has not convinced the 17-year-olds that the subjects are useful while the 13-year-olds still believe that these as-yet-unknown subjects are useful.

Mathematics as a Discipline. Some exercises dealt with the supposed sex-related stereotyping of mathematics as a male-dominated subject.

Table 13-22

Students' perceptions of mathematics as rule-oriented or process-oriented

Statement	Age	Percent Responding		
		Disagree	Undecided	Agree
Learning mathematics is mostly memorizing	13 17	33 40	18 14	48 45
There is always a rule to follow in solving mathematics problems	13 17	5 8	5 4	89 88
Doing mathematics requires lots of practice in following rules	13 17	12 8	11 12	77 80
Knowing how to solve a problem is as important as getting a solution	13 17	4 3	8 4	88 92
Knowing why an answer is correct is as important as getting the correct answer	13 17	4 3	7 4	88 93
Justifying the mathematical statements a person makes is an extremely important part of mathematics	13 17	4 5	31 28	65 68
Trial and error can often be used to solve a mathematics problem	13 17	13 10	31 19	56 70

The students across all age levels did not perceive mathematics as a male domain. Two-thirds of the 9-year-olds disagreed with both of the following statements: "Mathematics is more for boys than for girls" and "Mathematics is more for girls than for boys." Around 90 percent of both the 13- and 17-year-olds disagreed with both statements, with less than 5 percent of any age group agreeing with either statement. It seems clear that the students, particularly the older ones, do not in any sense perceive mathematics to be a male-dominated or female-dominated subject.

Several of the exercises dealt with 13- and 17-year-olds' perceptions of mathematics as either rule-oriented or process-oriented; results are summarized in Table 13-22. Although the students appear to hold divided opinions about whether learning mathematics is mostly memorizing, they almost unanimously felt that mathematics is very much rule based. Just as strongly, however, they felt that knowledge of process is as important as getting answers. These views seem almost contradictory, but both age groups have remarkably similar responses to the statements. One statement that did generate some disagreement between age groups was, "Trial and error can often be used to solve a mathematics problem." If most students really believe that doing mathematics requires following rules, then it seems that more students would disagree that trial and error could be used to solve problems.

It is interesting to speculate about students' reasons for responding to these particular questions as they did. Do they really feel what they said, or were they trying to give answers they thought were correct? The students probably did not consider their responses contradictory in light of their experiences with mathematics. For the most part, their mathematics has been oriented toward computation-type activities for which there *is* always a rule to follow; and, in order to be successful, one needs to practice following the rules. Students have also heard, however, that they must understand *why* rules work. Whether they really believe that or are merely paying lip service to a frequently espoused view is impossible to determine.

The results of the NAEP attitude assessment in mathematics represent a rich source of information about students' perceptions of mathematics and themselves as learners of mathematics, despite the fact that there might possibly be some legitimate questions that could be raised about the validity of the data. There have been few parallel opportunities to gather information about students' feelings about mathematics on such a large-scale and representative sample as that used by NAEP. Thus, these results represent a significant contribution to the literature of mathematics education and provide a unique perspective of the extent to which mathematics programs are developing a very basic skill, namely, favorable attitudes toward the subject.

CONCLUSIONS

One drawback to a list of basic skills is that it gives the impression that the listed items are separate, isolated skills. The National Assessment results clearly demonstrate how interdependent basic skills are with one another. Estimation and problem solving are only possible if students have a good understanding of the basic number, measurement, geometry, or probability concepts involved in the estimate or problem situation. Similarly, success in computation or measurement includes the ability to decide whether a result is reasonable or not.

The NAEP results indicate that students have generally learned the routine skills in most content areas. They are reasonably proficient at performing simple calculations, making measurements, and identifying basic geometric shapes. Where the greatest difficulty occurs is in the higher-order applications of these skills—in solving problems, making estimates, deciding on the reasonableness of a result, or applying knowledge of basic skills to everyday situations.

Information obtained from the attitude exercises suggests that mathematics is perceived as essential by all age groups. We need to take advantage of students' respect for mathematics and insure that sufficient emphasis is placed on all aspects of basic skills. The NAEP results do not provide final answers as to what direction the mathematics curriculum should take in the future, but they do document that some redirection of mathematics programs is needed, and they provide some baseline data about students' abilities that should be taken into account when making these decisions.

REFERENCES

Advisory Panel on the Scholastic Aptitude Test Score Decline. *On Further Examination.* New York: College Entrance Examination Board, 1977.

Carpenter, Thomas P.; Corbitt, Mary K.; Kepner, Henry S., Jr.; Lindquist, Mary M.; and Reys, Robert E. "Results and Implications of the Second NAEP Mathematics Assessment: Elementary School." *Arithmetic Teacher* 27 (April 1980): 10–12, 44–47(a).

Carpenter, Thomas P.; Corbitt, Mary K.; Kepner, Henry S., Jr.; Lindquist, Mary M.; and Reys, Robert E. "Results of the Second NAEP Mathematics Assessment: Secondary School." *Mathematics Teacher* 73 (May 1980): 329–39(b).

Carpenter, Thomas P.; Corbitt, Mary K.; Kepner, Henry S., Jr.; Lindquist, Mary M.; and Reys, Robert E. *Results and Implications of the Second NAEP Mathematics Assessment.* Reston, Va.: NCTM, in press.

Committee on Basic Mathematical Competencies and Skills. "Mathematical Competencies and Skills Essential for Enlightened Citizens." *Mathematics Teacher* 65 (November 1972): 671–77.

Commission on Post-War Plans of the NCTM."Guidance Report." *Mathematics Teacher* 40 (November 1947): 315–39.

National Assessment of Educational Progress. *Mathematics Objectives: Second Assessment.* Denver, Colo.: NAEP, 1978.

National Assessment of Educational Progress. *Changes in Mathematical Achievement, 1973–78.* Report no. 09-MA-01. Washington, D.C.: U.S. Government Printing Office, 1979(a).

National Assessment of Educational Progress. *Mathematical Applications.* Report no. 09-MA-03. Washington, D.C.: U.S. Government Printing Office, 1979(b).

National Assessment of Educational Progress. *Mathematical Skills and Knowledge.* Report no. 09-MA-02. Washington, D.C.: U.S. Government Printing Office, 1979(c).

National Assessment of Educational Progress. *Mathematical Understanding.* Report no. 09-MA-04. Washington, D.C.: U.S. Government Printing Office, 1979(d).

National Council of Supervisors of Mathematics. "Position Paper on Basic Skills." *Mathematics Teacher* 71 (February 1978): 147–52.

14. Recommendations for School Mathematics Programs of the 1980s

Shirley A. Hill

In April 1980 the National Council of Teachers of Mathematics (NCTM) presented a set of broadly ranging policy recommendations for school mathematics to the profession and to the public (National Council of Teachers of Mathematics 1980). The document has been called an agenda for a decade of action to improve mathematics learning and teaching. It dedicates the NCTM, an organization of over 75,000 members, to leadership in a coordinated program of planned positive change. Just as significantly, it calls upon all interested sectors of society to be partners in this effort—teachers, students, parents, school administrators, and other professional educators, school boards, teacher educators, and college mathematics teachers, legislators, government agencies, business and union leaders, and the public. This, it must be admitted, is a formidable and even idealistic endeavor. Nevertheless, the experience of recent decades has made it clear that the factors and pressures that determine curricula are complex and far-reaching and that the process of change must involve support and coordinated action on the part of many groups, not just the professional educators.

ASSUMPTIONS

It has been said that "education is everybody's business." The proper education of its young is one way a society ensures its future. In a democratic society, the public has a valid and legitimate role in determining educational goals and policy. But for such a philosophy to have beneficial results, public opinion, as interpreted by the people's representatives, must be well informed.

The responsibility of professional organizations whose members represent expertise in their appropriate areas is to inform the public and to make reasoned and responsible recommendations about educational goals and specific curricular policy. They should also provide explicit guidance to the education profession itself in the design of instructional programs that will advance the achievement of recommended goals and objectives. Judgments should proceed from the special knowledge and experience of the professional but be grounded in reality and responsible to the legitimate concerns of all sectors of society. Thus, professional recommendations need a substantial data base.

THE DATA BASE

The NCTM recommendations are probably unique with respect to the extensiveness of the recent and relevant information that provided its baseline. In 1975 a National Advisory Committee on Mathematical Education (NACOME), appointed by the Conference Board of the Mathematical Sciences and funded by the National Science Foundation, had pulled together available data for its report, *Overview and Analysis of School Mathematics, Grades K-12* (National Advisory Committee on Mathematical Education 1975). The report served as a benchmark, but it decried the serious paucity of information about what actually happens in mathematics classrooms. A survey by the committee strongly suggested that classroom practice had not undergone to any appreciable extent the profound changes suggested by the excitement of proposed curricular reform in the 1960s.

During the last half of the 1970s, a series of three status studies funded by the National Science Foundation helped to make up the deficiency in information about classroom practice. The NCTM appointed a team to do interpretive reports of these studies, and those reports helped provide a realistic base from which the policy recommendations could proceed.

There certainly has been no lack of data concerning test results in the last decade. These are the data that attract the greatest interest among the popular media and so contribute disproportionately to public opinion about education. At the national level and from the broadest perspective, the most useful assessment data were provided by the National Assessment of Educational Progress (NAEP). The results of the NAEP second assessment of mathematics achievement were released in 1979 (National Assessment of Educational Progress 1979). They received wide coverage in the national media and were extremely timely from the standpoint of contributing to the data base for the NCTM recommendations. Again, a NCTM team analyzed the results and developed interpretive summaries.

Most professionals in mathematics education who were asked to comment on the import of the results agreed: while students maintained an acceptable level of skills in computation with whole numbers, their abilities to apply their skills to the solution of problems (routine or nonroutine) were woefully deficient. The belief was generally expressed that an excessive narrowing of the curriculum to rote skills, under the pressures of the back-to-basics movement, was largely responsible. It can scarcely be denied that skills of computation are of little value if one cannot use them to solve problems.

Data about present status, or what exists, permit realism. To be responsible, recommendations also require data about people's opinions and priorities. What, we must ask, ought to be done?

THE PRISM SURVEY

The NCTM, jointly with Ohio State University, received a grant from the National Science Foundation to conduct a survey of the preferences various populations held on a wide range of issues in mathematics education (National Council of Teachers of Mathematics 1980b). The project, Priorities in School Mathematics (PRISM), sampled opinions of professionals and lay persons (teachers, supervisors, administrators, teacher educators, mathematicians, school board members, and parents). These data were considered in the development of the NCTM recommendations.

A few of the salient conclusions from the survey will be reported and discussed here. The reader should be aware, however, that hundreds of items were covered, and these few selected cannot do justice to the massive set of data collected.

Over the entire list of objectives, the development of problem-solving ability was given highest priority by *all* populations and by a wide margin. In ranking a number of possible goals for a problem-solving emphasis, the first-ranked goal for *all* populations was "to develop methods of thinking and logical reasoning."

Populations differed somewhat on how an additional fifteen minutes a day in elementary school should be spent. Elementary teachers, supervisors, and teacher educators gave first priority to problem solving, while high school teachers, principals, and the lay sample placed problem solving second to "drill on basics."

Over 80 percent in all samples would support an increased emphasis on applications throughout the curriculum. A continued emphasis on basic skills was supported by all samples except the teacher educators and supervisors, with strongest support from the lay sample. Preferences ranged widely, however, with regard to the specific activities or content that should receive greatest attention.

Support for learning to read mathematics was fairly high in all samples, but apparently the word *reading* was interpreted broadly because the highest support (95 percent) was given to the goal of learning to read graphs and data.

Supervisors and teacher educators are more favorable (75 percent) to increased use of calculators, but less than 40 percent of *any* sample would *decrease* the emphasis. It appears that the majority would have calculators used, but there is considerable variance in the particular ways and activities of preferred usage.

Strong support was given (75 percent of professional samples and 80 percent of the lay sample) for increased emphasis on the "use of computers and other technology" in the 1980s. Again, however, there is not a clear-cut blueprint for the specific components of such an emphasis. But on an item concerning preferences in the development of new secondary materials, all except the lay sample gave top priority to computer literacy, and the lay sample ranked computer literacy second after algebra. In general, recognition of the need to learn about computers is clear throughout the survey. In all samples, there was more disagreement than agreement on the statement, "Courses about computers should be strictly elective."

The need for more, and more varied, instructional resources was evidenced in all teacher samples (elementary, secondary, two-year college, and four-year college). Strong support was reported for use of experiments, manipulative materials, and measuring devices. Moderate to strong support was given to having students work in small groups.

Increased emphasis on homework was supported by 82 percent of the college samples but only 50 to 60 percent of the other samples. Principals and school board members strongly supported (84 and 91 percent, respectively) text materials with daily homework problems, but support among parents was somewhat lower (67 percent).

The response by the lay sample to questions about requirements in high school mathematics indicates that such requirements may presently be lower than the general public deems desirable. Lay persons responded in these ways with respect to years of high school mathematics they would require:

For college-bound students: four years (47 percent); three years (36 percent); two years (13 percent); one year or less (3 percent).

For graduation requirements for all students: four years (15 percent); three years (25 percent); two years (47 percent); one year (12 percent); none (1 percent).

The average suggested by the *lay* sample for all students was 2.4 years.

While samples split in their support for a full-year algebra course for all high school students (80 percent of mathematicians, 24 percent of supervisors), there was overall support of 70 percent for the statement, "Different algebra courses should be offered for students with different interests and abilities."

On items concerning special attention to special groups of students, all samples agreed to a high priority for students with learning and other handicaps. Samples were lukewarm or decidedly mixed with respect to emphasis on other specific groups of students. An exception was the fairly strong support for increasing attention to gifted students. Overall, 75 percent believed there should be more emphasis on the gifted, with the least support from teachers in two-year colleges and mathematicians (60 and 70 percent, respectively).

On methods of attacking problems in mathematics education, highest rank was accorded the in-service education of teachers, with preservice education second. The ranking, in order of seriousness, on the general problems of the classroom teacher was: (1) unmotivated students, (2) reading difficulties, (3) classroom discipline, (4) no commitment to homework, (5) lowering of school academic standards, and (6) irregular attendance of students.

THE RECOMMENDATIONS

These data and much more were carefully analyzed by committees in the process of developing the recommendations made in *An Agenda for Action: Recommendations for School Mathematics of the 1980s* (National Council of Teachers of Mathematics 1980a). It should be said, however, that while such data help to insure realism and responsibility, recommendations, whether of fact or opinion, do not ensue directly from such information. Professional experience and knowledge play their vital roles in generating what, in the opinion of professionals in school mathematics instruction, is considered in the best interests for the education of today's students in tomorrow's world.

An Agenda for Action is organized into eight major categories with more specific recommendations for action following each category. The eight major recommendations are:

1. that problem solving be the focus of school mathematics in the 1980s
2. that basic skills in mathematics be defined to encompass more than computational facility
3. that mathematics programs take full advantage of the power of calculators and computers at all grade levels
4. that stringent standards of both effectiveness and efficiency be applied to the teaching of mathematics
5. that the success of mathematics programs and student learning be evaluated by a wider range of measures than conventional testing
6. that more mathematics study be required for all students and a flexible curriculum with a greater range of options be designed to accommodate the diverse needs of the student population
7. that mathematics teachers demand of themselves and their colleagues a high level of professionalism
8. that public support for mathematics instruction be raised to a level commensurate with the importance of mathematical understanding to individuals and society.

The first four recommendations are closely interrelated and, in fact, address different aspects of a new curricular emphasis. They anticipate a future in which the demands and needs in mathematical skill and knowledge will be different than they have been. They are a response to the rapid expansion of mathematical knowledge, increasing uses and application of mathematical methods, and the dizzying pace at which the technological tools that serve problem solving have developed. The relationship is complex. The remarkable capacities of new microtechnology condition the skills and abilities most likely to be advantageous in future decades. Conversely, an orientation toward the primacy of problem-solving ability as a goal demands tools with capabilities of flexibility and interaction not provided by previous instructional media.

The implications are myriad, but some are both clear and urgent. Our notion of what is basic or essential for all students in mathematics must include far more than the restricted computational facility often cited as "the basics." Assessments have indicated that the schools do a better job on the development of whole number computation than on any of the several other areas that should also be considered basic. Yet many school programs have increased the intense focus on computation almost exclusively of a serious concentration on the other basics.

All students need skill in measurement and skill in applying concepts and techniques of percent and ratio to a variety of everyday problems. In a world in which every citizen is constantly bombarded with statistical data, often designed to persuade or influence, the ability to deal intelligently with quantitative information is essential.

Estimation is a vital skill that can be taught, and school instruction can encourage the habit of mind that consistently evaluates the reasonableness of a proposed answer or result. Students also need to learn to make judgments about the sensible choices among skills and tools: when an estimate is sufficient, when mental computation is a useful skill, when paper-and-pencil calculation is practical, and when conditions dictate the use of a calculator.

Perceptual capacity to deal with relationships among forms in space has long been known to be a factor in mathematical problem-solving ability. Yet school curricula have not had a systematic program that helps the development of such capacities. Activities with physical materials, used systematically, can improve spatial concepts and abilities. Such abilities are at the least useful, and they are essential in many occupations. Closely related are problem-solving skills that involve diagramming, graphing, and mental imagery. A repertoire of heuristic strategies, including physical and graphic modeling and imagery, should be considered basic.

An agenda for the immediate and distant future in mathematics instruction must include careful planning for three objectives: (1) the full and appropriate use of technological tools, principally computers and calculators, *for* developing problem-solving ability; (2) teaching students *how* and *when* to use these tools; (3) and teaching *about* the uses and limitations of technological tools. Thus the NCTM recommendations address the need for software that takes full advantage of the potential of microcomputers to reach learning objectives, the need to include calculators and computers at all school levels, and the need for computer literacy for all students (and, indeed, all citizens).

The fourth recommendation is implied by the interrelationships of the first three. To effect the changes recommended, a rethinking and perhaps a significant reorganization of the school program is needed. At the least, more time should be spent on mathematics instruction and study than is currently the norm. The heart of the recommendations in the fourth category is the recognition of how precious instructional time is and the significant impact that the amount of instructional time and the efficient management of that time have on learning. In this context there are three central assumptions, and the acceptance of these assumptions implies a need for serious, if not radical, change in classroom priorities.

The first assumption is that as mathematics and its uses have grown rapidly, the need for mathematical skill has become greater and embraces more and more job, consumer, and citizen roles. Thus, if the traditional presumption that most people need to know arithmetic and perhaps a little algebra (with the luxury of geometry to "teach one to reason") and all else is for specialists was ever valid, it certainly is not so today. Most students need more of that precious commodity, instructional time, devoted to mathematics.

The second assumption reinforces that conclusion. The *Agenda* makes the following statement: "Higher-order skills in problem solving require more time to learn than the lower-order, narrowly mechanistic skills (p. 12)." Certainly an environment proposed by the *Agenda*, that is, an exploratory, experimental, questioning, estimating, testing environment in which a wide variety of heuristic approaches, materials, and models are experienced, occupies fairly extensive chunks of time and requires careful preplanning and good classroom management.

The third assumption is that needs for skills change over time; that as new tools and methods are developed, some formerly essential skills become obsolete. But it is very difficult to winnow out the "no-longer-basic" from the school curriculum. Tradition probably keeps more content in the curriculum longer more than any careful assessment of what will be applicable in the future.

The difficulties of winnowing out obsolete material are exacerbated by the obvious fact that skills do not achieve obsolescence overnight; rather there is a long transition as society adjusts to changing demands. The art is in recognizing and predicting trends and social change. These uncertainties do not eradicate the necessity for priority setting, however, and policy makers and curriculum developers must balance finite classroom time against changing educational needs. Mathematics skills and knowledge must be judged in terms of a "time-effective" criterion—weighing the time required for mastery against future applicability in realistic situations. An example might be to ask if the large investment of time spent on the long-division algorithm with multiple-digit divisors can be justified in an age of electronic aids, especially if it is at the expense of instruction in problem solving.

Recommendation five reiterates a position on evaluation long taken by many professional educators. This position, while not new, needs to be repeated and stressed as school officials and the public increasingly rely solely on tests in judging the effectiveness of programs and the success of individual learners. Test results have become nearly synonymous with success or failure, yet the limitations and abuses of tests are widely known.

The NCTM *Agenda* sees tests as only one instrument for evaluation and one that should be used with care. It also points out that a serious acceptance of problem-solving ability as a school objective creates a demand for new techniques in assessment. It challenges the present testing models as inadequate for the task.

The sixth recommendation contains a recommended action certain to be controversial. "At least three years of mathematics should be required in grades nine through twelve (p. 20)." It is interesting that such a recommendation may find more immediate support among the general public (witness the PRISM survey and the Gallup Poll on public attitudes toward education) than among mathematics teachers, who correctly foresee the enormous difficulties in implementation. Nonetheless, if one argues for the status quo—typically a requirement of one year of mathematics in grades nine through twelve—one must argue that for about half the student population, sufficient mathematics has been learned by age fourteen. The other possible assumption, that students will recognize their best future interests and *elect* more mathematics is demonstrably invalid. (It is interesting that policy makers apparently do not assume that students can be relied upon to elect English courses in their best interests.)

The Mathematical Association of America, in a conference report (PRIME 80), has recommended three years of high school mathematics for the college-bound student. Unless we return to a strict tracking system of academic and nonacademic students (which few would advocate), we should keep options open for some form of college or technical education for most students. Thus for the majority, it can be argued that three years will be advantageous if not essential.

The *Agenda* recommendation does not rest on the "just-in-case" argument of keeping doors open. It clearly assumes that almost all students can profit from three years of good mathematical study in high school, though it emphasizes that this will not be the same set of courses for all. In fact, part of the recommendation acknowledges that curricula must be made more flexible and diverse to accommodate varying needs, interests, and abilities. This is obviously a serious challenge to the mathematics education profession itself.

It is recognized that implementation is made even more difficult by the increasing shortages in qualified mathematics teachers. But this state of affairs does not obviate the need of both the individual and society for mathematical ability and certainly does not excuse a professional organization of mathematics teachers from the responsibility to make the public aware of both problems. In fact, such a recommendation puts the responsibility for a solution precisely where it belongs—with society and the public representatives. Is society willing to pay the price, whatever that price is, to have its children learn sufficient mathematics for their future well-being? The profession of mathematics educators or the profession of educators cannot alone solve these problems.

The NCTM *Agenda* speaks to this issue in its last two recommendation categories. One is a call to renewed commitment to professionalism. This includes the obligation of individuals to remain current, continue learning, and maintain professional priorities as well as the obligation of school administrations and school boards to assist and support teachers in their professional efforts and activities.

This assistance links up with the final recommendation—that fundamental to achieving all the other recommended actions is public support. In a nutshell, all the recommended actions in the last category and, indeed, the ultimate purpose of policy recommendations of this kind are summed up in this paragraph at the end of the *Agenda*:

> The professional community and society share a common goal: to bring all citizens to the full realization of their mathematical capacity. This is a complex and delicate task, and it requires the commitment and cooperation of all segments of society, not just the school, parents, and teachers. (p. 28)

REFERENCES

Mathematical Association of America. *PRIME—80: Proceedings of a Conference on Prospects in Mathematics Education in the 1980s.* Washington, D.C.: Mathematical Association of America, 1978.

National Advisory Committee on Mathematical Education (NACOME). *Overview and Analysis of School Mathematics, Grades K-12.* Conference Board of the Mathematical Sciences. Reston, Va.: NCTM, 1975.

National Assessment of Educational Progress. *Reports of the Second Assessment of Mathematics.* Denver, Colo.: Education Commission of the States, 1979. (Available from NAEP, Suite 700, 1860 Lincoln St., Denver, Colo., 80295.)

National Council of Teachers of Mathematics. *An Agenda for Action: Recommendations for School Mathematics of the 1980s.* Reston, Va.: NCTM, 1980(a).

National Council of Teachers of Mathematics. *A Study of Priorities in School Mathematics.* Columbus, Ohio: Ohio State University Center for Science and Mathematics Education, 1980(b).

Stake, Robert E., and Easley, Jack, eds. *Case Studies in Science Education.* Urbana, Illinois: University of Illinois, 1978.

Suydam, Marilyn N., and Osborne, Alan. *The Status of Pre-College Science, Mathematics, and Social Science Education, 1955-1975.* Volume II: Mathematics Education. Columbus, Ohio: Ohio State University Center for Science and Mathematics Education, 1977.

Weiss, Iris. *Report of the 1977 National Survey of Science, Mathematics, and Social Studies Education.* Research Triangle Park, N.C.: Research Triangle Institute, 1978.